Silver Educational Publishing
Published by Silver 8 Production LLC

ISBN 0-9743287-7-4

Additional Study Guides Available For National Board Exams From Silver Educational Publishing

National Board of Chiropractic Part I Study Guide: Key Review Questions and Answers

National Board of Chiropractic Part II Study Guide: Key Review Questions and Answers

National Board of Chiropractic Part III Study Guide: Key Review Questions and Answers with Explanations

National Board of Chiropractic Part IV Study Guide: Key Review Questions and Answers (Topics: Diagnostic Imaging) Volume 1

National Board of Chiropractic Physiotherapy Study Guide: Key Review Questions and Answers

SILVER EDUCATIONAL PUBLISHING
SEPBOOKS.COM

NATIONAL BOARD of CHIROPRACTIC* PART IV STUDY GUIDE: *Key Review Questions and Answers*

Topics

Case Management & Technique Practical

Volume 2

Dr. Patrick Leonardi

This is Volume 2 of a two volume series. Make sure to also purchase the following:
- NATIONAL BOARD OF CHIROPRACTIC PART IV STUDY GUIDE: KEY REVIEW QUESTIONS AND ANSWERS (TOPICS: DIAGNOSTIC IMAGING) VOLUME 1

Contents

Case Management

1. A 40-year-old truck driver has been diagnosed with Horner's syndrome. Recently, he had sustained a serious whiplash injury due to a car accident. Anhydrosis is present. Name two other clinical signs or symptoms observed in Horner's syndrome.

a. resting tremor
b. ptosis
c. miosis
d. blindness
e. migraine
f. tinnitus
g. vertigo
h. adiadokokinesia
i. cyanosis
j. tonsillitis

2. According to the previous question, name two causes for this condition.

a. vertigo
b. Pancoast tumor
c. prostate cancer
d. hyperglycemia
e. hypoglycemia
f. melanoma
g. Bell's Palsy
h. intention tremor
i. severe whiplash injury
j. scleroderma

3. A 70-year-old male bartender has had emphysema for over 5 years. Which sound on percussion will be assessed? Also, name a finding associated with this condition on x-ray.

a. resonant
b. hyperresonant
c. dull
d. tympanic
e. flat
f. blebs
g. plasmacytoma
h. hematoma
i. multiple myeloma
j. Haygarth's nodes

4. A 40-year-old frequent restaurant patron has jaundice. She has some abdominal pain and nausea. Which is the diagnosis? Also, name another common symptom or sign for this condition.

a. hepatitis A
b. cholecystitis
c. hydrocephalus
d. SLE
e. subacute bacterial endocarditis
f. fever
g. tinnitus
h. alopecia
i. joint pain
j. wheezes

5. A 36-year-old female has intention tremor, optic neuritis and spastic

paralysis with remissions. Which is the diagnosis? Also, identify the cause of this condition.

a. Parkinson's
b. syringomyelia
c. multiple sclerosis
d. myasthenia gravis
e. multiple myeloma
f. loss of dopamine
g. myo-neural defect
h. demyelination of CNS
i. injury to cranial nerves
j. Pancoast tumor

6. A 36-year-old woman presents with ptosis, dysarthria and difficulty swallowing. Which auto-immune disease does this best represent?

a. rheumatoid arthritis
b. DJD
c. Parkinson's
d. myasthenia gravis
e. scleroderma
f. Wilson's disease
g. SLE
h. syringomyelia

7. According to the previous question, name a sign or symptom of this condition?

a. resting tremor
b. acro-osteolysis
c. loss of pain and temperature
d. weakness of muscles innervated by cranial nerves
e. ulnar deviation
f. unilateral loss of joint space
g. iron toxicity
h. butterfly rash

8. Which of the following is a cause of mitral stenosis? Also, name the best location near the heart for listening to this condition.

a. rheumatic fever
b. Raynaud's disease
c. thrombophlebitis
d. aortic aneurysm
e. multiple myeloma
f. base of heart
g. apex of heart
h. tricuspid area
i. right second intercoastal border
j. pulmonic area

9. A 42-year-old male had recently experienced deep pain that radiated toward the jaw, neck and arm. In addition, he was rushed to the emergency room and was diagnosed with a myocardial infarction. Which structure of the heart is usually predominately affected in this condition? Also, which lab test will be found to be elevated and is commonly used to assess myocardial infarctions?

a. left ventricle
b. right ventricle
c. left atrium
d. right atrium
e. HLA B-27
f. BUN
g. albumin/globulin ration
h. CPK
i. amylase
j. alpha-fetoprotein

10. Name two other profiles or laboratory tests used to evaluate MI?
a. lipase
b. T-3
c. T-4
d. SGOT
e. LDH
f. TSH
g. uric acid
h. alkaline phosphatase
i. A/G ratio
j. LE cell

11. A 24-year-old female has hepatitis B. Which of the following lab values will be increased in this condition? Also, name a sign or symptom of this condition.
a. CPK
b. LE cell
c. bilirubin
d. uric acid
e. LDH
f. jaundice
g. tinnitus
h. ptosis
i. hyperacusis
j. rebound phenomenon of Holmes

12. Range of motion of the shoulder is attempted on a 35-year-old male. However, this test could not be performed due to the patient feeling pain when attempting passive shoulder abduction. In addition, the Apprehension test is positive. What is the most likely diagnosis?
a. lateral epicondylitis
b. medial epicondylitis
c. radiohumeral bursitis
d. shoulder dislocation
e. thoracic outlet syndrome
f. costoclavicular syndrome
g. multiple sclerosis
h. cervical foraminal compression

13. According the previous question, which is another orthopedic test that can confirm your diagnosis?
a. Yergason's
b. Dugas' test
c. Roos test
d. Halstead maneuver
e. Reverse Bakody maneuver
f. Allen's test
g. Eden's test
h. Wright's test

14. A 21-year-old pitcher has a possible rotator cuff tear. Which orthopedic test is most appropriate to perform?

a. Dugas' test
b. Yergason's test
c. Apley's test
d. Drop Arm Test
e. Nachlas test
f. Bonnet's sign
g. supported Adam's test
h. Lewin Punch test

15. According to the previous question, Codman's sign is present. Which is the most likely diagnosis?

a. thoracic outlet syndrome
b. tennis elbow
c. biceps tendon instability
d. supraspinatus tear
e. teres minor tear
f. radiohumeral bursitis
g. subacromial bursitis
h. Keinbock's disease

16. Thoracic outlet syndrome is diagnosed in a 39-year-old waiter. Pick two tests from the following list that may confirm this diagnosis.

a. Spurling's test
b. Distraction test
c. Roos test
d. Apley's test
e. Yergason's test
f. Apprehension test
g. Dugas' test
h. Mill's test
i. Shrivel test
j. Halstead maneuver

17. A 59-year-old female teacher presents with a red flushed face. Her chief complaint is a headache with some dizziness. Her previous history does show occasional bleeding gums. Lab values reveal the following: HCT is 51%. Which is the diagnosis?

a. Reye's syndrome
b. reflex sympathetic dystrophy syndrome
c. thrombosis
d. peptic ulcer
e. SLE
f. scleroderma
g. polycythemia vera
h. pancreatitis

18. According to the previous question, what does this condition cause?

a. excessive red blood cells
b. encephalopathy
c. butterfly rash
d. MI
e. acid reflux
f. persistent arteriole vasospasm
g. polyarthritis
h. maculo-papular rash

19. According to question 17, which is the most appropriate treatment protocol?

a. upper cervical adjustment
b. ultrasound
c. warm whirlpool
d. ice packs
e. advise on proper nutrition
f. transverse friction massage
g. refer to medical doctor
h. trigger points

20. Name two conditions where Argyll Robertson pupil can be observed.

a. hyperthyroidism
b. myxedema
c. tertiary syphilis
d. Charcot's triad
e. Addison's disease
f. diabetes mellitus
g. multiple myeloma
h. hydrocele
i. spider angioma
j. tuberculosis

21. A 42-year-old carpenter has been diagnosed with frozen shoulder syndrome. Which of the following would be most appropriate in the treatment of this condition?

a. Blount's exercises
b. McKenzie exercises
c. Codman's exercises
d. Frenkel's exercises
e. Philadelphia collar
f. corset
g. Kegel exercises
h. spinal traction

22. Rugger jersey spine and calcinosis are apparent on radiographs. Lab values indicate high calcium and alkaline phosphatase levels. Which is the diagnosis?

a. Addison's disease
b. tuberculosis
c. multiple myeloma
d. AS
e. psoriatic arthritis
f. Reiter's syndrome
g. hyperparathyroidism
h. rickets

23. According to the previous question, name two signs or symptoms for this condition.

a. constipation
b. arthritis
c. nervousness
d. Sprengel's deformity
e. epitransverse
f. platybasia
g. saber shin
h. hematoma
i. sluggishness
j. cafe-au-lait pigmentation

24. A 39-year-old office worker has compression of the posterior tibial nerve. When working out at the gym, jogging on the treadmill is quite painful. Which is the most likely diagnosis?

a. chondromalacia patellae
b. Volkmann's contracture
c. tarsal tunnel syndrome
d. shin splints
e. Baker's cyst
f. grade 1 ankle sprain
g. tear of the Achilles tendon
h. thrombophlebitis

25. A 45-year-old male smoker complains of pain in the legs when walking or running. However, the pain is relieved with rest. Which is the diagnosis?

a. vascular claudication
b. neurogenic claudication
c. thrombophlebitis
d. sciatica
e. spinal stenosis
f. thoracic outlet syndrome
g. fibrous dysplasia
h. neurofibromatosis

26. A 17-year-old male has anterior wedged vertebrae in the thoracic spine. Disc spaces are decreased. Which is the most probable condition?

a. hypoparathryoidism
b. AS
c. Reiter's syndrome
d. psoriasis
e. Scheuermann's disease
f. Sudecks atrophy
g. Klippel-Feil syndrome
h. scurvy

27. Name two treatments that are appropriate for chondromalacia patellae.

a. strengthen vastus medialis
b. strengthen hamstrings
c. Frenkel's exercises
d. Blount's exercises
e. Kegel contractions
f. corset
g. McKenzie exercises
h. halo-body jacket

28. A 38-year-old female presents with general nervousness and an extreme amount of energy. The patient complains of hair loss and wonders if she needs some nutritional supplement. Lid retraction is evident along with exophthalmos. Which is another clinical sign or symptom of this condition?

a. eye puffiness
b. constipation
c. drowsiness
d. weight loss
e. menorrhagia
f. retinoblastoma
g. ventricular septal defect
h. hiatal hernia

29. According to the previous question, what condition does this describe?

a. myxedema
b. Sudecks atrophy
c. osteopenia
d. polymyositis
e. stasis dermatitis
f. pulmonary embolism
g. Raynaud's phenomenon
h. hyperthyroidism

30. Name two other signs or symptoms associated with hyperthyroidism.
a. goiter
b. bradycardia
c. coarse hair
d. cluster headaches
e. kidney stones
f. Kerr's sign
g. double vision
h. decreased metabolism
i. scleroderma
j. butterfly rash

31. The posterior tibia reflex is diminished in a 41-year-old male. There is pain over the SI joint down to the lateral leg to the web of the big toe. In addition, there is limited dorsiflexion of the foot. The most probable cause is which of the following?
a. cauda equina syndrome
b. L3 disc herniation
c. osteoporosis
d. scleroderma
e. psoriasis
f. multiple myeloma
g. osteoma
h. multiple sclerosis

32. A 72-year-old female complains of substernal pain while walking. The pain goes away in 5 minutes after walking. She safely takes nitroglycerin to relieve the pain. The diagnosis is which of the following?
a. stroke
b. MI
c. cervical radiculopathy
d. multiple myeloma
e. osteoporosis
f. thrombophlebitis
g. angina pectoralis
h. lymphosarcoma

33. A 35-year-old male patient complains of a stabbing pain behind the left eye. The patient has an alcohol content of approximately 2 beers per day. In addition, he frequently wakes up with a headache. Which are two possible causes for this condition?
a. hypertension headache
b. aneurysm headache
c. cluster headache
d. multiple myeloma
e. AS
f. neurogenic claudication
g. DJD
h. osteoporosis
i. Sudecks atrophy
j. Raynaud's phenomenon

34. Name two signs or symptoms related to aneurysm headaches.

a. double vision
b. weight loss
c. coarse hair
d. nervousness
e. increase in metabolism
f. ptosis
g. nasal polpys
h. malocclusion
i. Keyser-Fleischer ring
j. ectropion

35. A 25-year-old female professional athlete complains of severe pain around both eyes. She complains that she sees flickering lights in her vision that sometimes occur before the pain. The cephalgia lasts from 15-24 hours. Which is the diagnosis?

a. cluster headache
b. tension headache
c. brain tumor
d. iritis
e. classic migraine
f. common migraine
g. hypertensive headache
h. malingering

36. According to the previous description, what would be an appropriate question to further confirm this diagnosis?

a. Do you have high blood pressure?
b. Do you drink a lot of alcohol?
c. Did your mother or siblings have migraines?
d. Do you exercise a lot?
e. Have you ever been diagnosed with cancer?
f. Did you ever have syphilis?

37. Name two factors that can bring on a cluster headache?

a. chocolate
b. alcohol
c. tomatoes
d. potatoes
e. high altitudes
f. scotomas
g. garlic
h. onion
i. whole wheat flour
j. selenium

38. A 55-year-old male college professor complains of headaches that have become progressively worse over the last 3 months in intensity and duration. The pain is near the left temporal bone. In fact, the pain is worse in the morning. Which is the most probable diagnosis?

a. common migraine
b. classic migraine
c. TMJ headache
d. cluster headache
e. Buerger's disease
f. bruxism
g. brain tumor
h. tension headache

39. According to the previous question, what are two clinical findings for this condition?

a. seizures
b. aura
c. jaw clicking
d. malocclusion
e. low blood sugar
f. projectile vomiting
g. runny nose
h. photophobia
i. scotomas
j. obesity

40. According to question 38, what is the proper treatment protocol for this condition?

a. adjust C1
b. tell patient to avoid alcohol
c. treat as medical emergency
d. tell patient to exercise more
e. adjust tempor-mandibular joint
f. tell patient to avoid chocolate
g. refer to dentist
h. adjust C2 for 3 times per week for 6 weeks

41. A 52-year-old author complains of jabbing pain around the right ear when eating. She has noticed moderate weight loss. In addition, her eye sight has become progressively worse. Blood pressure is 165/105 mm Hg. Which of the following is the diagnosis?

a. temporal arteritis
b. common migraine
c. classic migraine
d. thrombophlebitis
e. hypoglycemia
f. multiple myeloma
g. osteoporosis
h. tension headache

42. According to the previous question, what are two conditions that can result from this diagnosis?

a. jaw clicking
b. pathological bone fracture
c. punched-out lesions
d. raindrop skull
e. blindness
f. diabetes mellitus type II
g. pancreatitis
h. stroke
i. psoriatic arthritis
j. TMJ

43. A 56-year-old woman has intense, shock-like pain around the jaw and mouth areas. The pain started after chewing gum. Which is the most probable condition?

a. hunger headache
b. tic douloureux
c. Bell's Palsy
d. multiple myeloma
e. osteoma
f. stroke
g. scleroderma
h. syphilis

44. Which part of the body causes the above condition described?

a. inflammation of cranial nerve 7
b. osteoporosis
c. inflammation of cranial nerve 5
d. Reiter's syndrome
e. rheumatoid arthritis
f. anemia
g. low glucose levels
h. obesity

45. A 71-year-old retired school teacher presents with a mask-like expression. Resting tremor and drooling are also apparent. Which is the condition?

a. Huntington's chorea
b. multiple sclerosis
c. osteopenia
d. Parkinson's disease
e. dacryocystitis
f. squamous cell carcinoma
g. pulmonary edema
h. Sjogren's syndrome

46. Name another clinical sign of the above condition.

a. fever
b. atelectasis
c. intention tremor
d. metastasis
e. pathological fracture
f. forward flexion of neck and trunk
g. nasal congestion
h. nasal polyps

47. A 2-year-old infant presents with a stridulous cough coupled with dyspnea. Which is the diagnosis?

a. croup
b. sarcoidosis
c. pneumothorax
d. pleurisy
e. leukemia
f. anemia
g. squamous cell carcinoma
h. Hodgkin's disease

48. A 74-year-old patient has chills, blood filled sputum, fever of 105° F, and chest pain. Which is the most likely diagnosis?

a. sarcoidosis
b. croup
c. bronchitis
d. pneumonia
e. atelectasis
f. asthma
g. fibroma
h. hamartoma

49. A 40-year-old female has a palpable, hot vein in the calf. Which of the following best represents this description?

a. neurogenic claudication
b. thrombophlebitis
c. appendicitis
d. multiple myeloma
e. volvulus
f. Peutz-Jegher's syndrome
g. diverticulum
h. Turcot's syndrome

50. Which of the following would you expect to find positive for the previous condition described?

a. Minor's sign
b. Beever's sign
c. Homan's sign
d. Bikele's sign
e. Guilland's sign
f. Adam's procedure
g. Hoover's sign
h. Schepelmann's sign

51. Which two of the following would be positive in a multiple sclerosis patient?

a. Spinal percussion test
b. Rust's sign
c. Beever's sign
d. Homan's sign
e. Forestier's Bowstring sign
f. Lewin supine test
g. Ballottement test
h. Brudzinski sign
i. Minor's sign
j. Guilland's sign

52. A 45-year-old postal worker has been diagnosed with Wegener's granulomatosis.. ESR is elevated also. In addition, moderate hematuria is present. Name two signs or symptoms present in this condition.

a. malaise
b. genital warts
c. syphilis
d. pulmonary congestion
e. appendicitis
f. acne
g. butterfly rash
h. acromegaly
i. Cullen's sign
j. positive Libman's sign

53. A 48-year-old female pastry chef presents with fever, tachycardia and tenderness in the right upper quadrant of the abdomen. In addition, pain is

felt at the right shoulder and interscapular area. Gallstones are suspected. The patient is also 35 pounds overweight. Which is the most probable condition?

a. cirrhosis
b. cholecystitis
c. scleroderma
d. rheumatoid arthritis
e. fibromyalgia
f. multiple myeloma
g. anemia
h. pancreatitis

54. According to the previous question which three tests can confirm this diagnosis?

a. Complete Blood Count
b. HLA-B27
c. HDL
d. glucose tolerance test
e. ultrasound
f. gamma-glutamyl transferase
g. cholecystography
h. human chorionic gonadotropin
i. Reed Sternberg cell
j. mean corpuscular hemoglobin

55. According to question 53, identify three signs or symptoms related to this condition.

a. steatorrhea
b. blood streaked stools
c. osteoma
d. pathological fractures
e. punched out lesions
f. more than 10 painful, tender spots
g. blue sclera
h. osteoporosis
i. flatulence
j. pes planus

56. A 52-year-old male has a positive Cullen's sign and ecchymosis of the flanks. Pain is constant along the epigastric area. The patient's level of alcohol is at least 2 drinks per day. Identify the most likely condition.

a. cholecystitis
b. pancreatitis
c. scleroderma
d. osteosarcoma
e. hamartoma
f. cholelithiasis
g. fibromyalgia
h. sarcoidosis

57. According to the previous question, what condition may cause this diagnosis?

a. rheumatoid arthritis
b. DJD
c. alcoholism
d. anemia
e. AS
f. hemorrhage
g. Haygarth's nodes
h. leukemia

58. According to question 56, name two clinical signs of this condition.

a. disc protrusion
b. anterior osteophyte
c. intercostal neuritis
d. fertile
e. hip dislocation
f. abdominal distention
g. low blood pressure
h. Scandinavian descent
i. Raynaud's phenomenon
j. metastasis

59. A 43-year-old woman presents with swan neck deformity and rat bite erosions in the phalanges. Lab values reveal elevated alkaline phosphatase and ESR. Which is the most probable condition?

a. RA
b. DJD
c. scleroderma
d. Reiter's syndrome
e. sarcoidosis
f. psoriatic arthritis
g. osteosarcoma
h. fibromyalgia

60. Name two other findings consistent with the previous condition described in question 59.

a. osteophytes
b. juxta-articular osteoporosis
c. non-uniform loss of joint space
d. geodes
e. pseudocysts
f. Salter Harris fracture type 1
g. diarrhea
h. Cullen's sign
i. calcinosis
j. urethritis

61. Which condition is most closely associated with rheumatoid arthritis?

a. multiple myeloma
b. Kawasaki's disease
c. Wegener's granulomatosis
d. hypoparathyroidism
e. Felty's syndrome
f. multiple myeloma
g. sickle-cell anemia
h. rickets

62. According to the answer in question 61, identify two clinical signs or symptoms of this condition.

a. splenomegaly
b. osteophytes
c. positive Schepelmann's sign
d. positive Homan's sign
e. resting tremor
f. leukopenia
g. festinating gait
h. Charcot's triad
i. scotomas
j. pigeon breast deformity

63. A 32-year-old woman has been diagnosed with syringomyelia. What

are two clincial features of this condition?

a. resting tremor
b. intention tremor
c. stiffness in the shoulders & back
d. capelike sensory insufficiency on upper torso
e. osteophytes
f. mask-like face
g. positive Lhermitte's test
h. waddling gait
i. positive Dawbarn's sign
j. osteomyelitis

64. A 49-year-old male teacher presents with multiple spider angiomas on the chest and lower extremities. Which are two possible causes for this condition?

a. pustules
b. petechia
c. psoriasis
d. vitamin B deficiency
e. syphilis
f. liver disease
g. trichinosis
h. anemia
i. Lyme disease
j. eczema

65. A 16-year-old high school baseball player presents with fatigue, lymphadenopathy, high fever and a sore throat. Which is the diagnosis?

a. Kaposi's sarcoma
b. toxoplasmosis
c. Hodgkin's disease
d. neurogenic claudication
e. juvenile rheumatoid arthritis
f. scleroderma
g. mononucleosis
h. croup

66. According to the previous question, which two tests would you recommend to confirm this diagnosis?

a. Paul Bunnell
b. SGOT
c. Downey Cell
d. ELISA
e. glucose tolerance test
f. radiographs of the hand
g. gamma-glutamyl-transpeptidase
h. HCG
i. LE cell
j. LDH

67. According to question 65, which is responsible for this condition?

a. Reed Sternberg cell
b. HIV virus
c. history of smoking
d. lack of exercise
e. brain tumor
f. Epstein Barr Virus
g. Toxoplasma gondii
h. neurotoxins

68. A 7-year-old male presents with several, severe wheezing occurrences with dyspnea. Which is the usual percussive note for this condition?

a. dull
b. rales
c. resonant
d. flat
e. hyperresonant
f. egophony
g. rhonchi
h. friction rub

69. According to the description from the previous question, what is this condition?

a. bronchitis
b. emphysema
c. pleurisy
d. pneumonia
e. asthma
f. bronchiectasis
g. atelectasis
h. cystic fibrosis

70. A 19-year-old female presents with painful joints, facial rash and sensitivity to sunlight. Lab test indicate +ANA. Which is the most likely diagnosis?

a. cystic fibrosis
b. systemic lupus erythematosus
c. emphysema
d. syphilis
e. herpes simplex
f. scleroderma
g. ankylosing spondylitis
h. temporal arteritis

71. Based on the previous question, name two other characteristics of the diagnosis?

a. kidney dysfunction
b. mouth sores
c. fever
d. hypothyroidism
e. diabetes insipidus
f. dagger sign
g. barrel chest
h. shiny corner sign
i. podagra
j. goiter

72. Which condition of bone does SLE most closely resemble?

a. scleroderma
b. osteosarcoma
c. osteoma
d. temporal arteritis
e. rheumatoid arthritis
f. multiple myeloma
g. DJD
h. rickets

73. Sciatica is suspected in a 35-year-old male. Which of the following would be positive? (Pick three choices.)

a. Lewin Standing test
b. Hibb's test
c. Gaenslen's test
d. Iliac compression test
e. Ober's test
f. Sicard's sign
g. Lindner's sign
h. Wartenberg's sign
i. Braclet test
j. Turyn's sign

74. A 45-year-old female has tenosynovitis of the thumb. Which of the following would most likely be positive?

a. Finsterer's sign
b. Finkelstein's test
c. Tinel's sign
d. Kaplan's sign
e. Dawbarn's sign
f. Mill's test
g. Roos test
h. Bakody sign

75. A 45-year-old personal trainer has microaneurysms in the eye coupled with acetone breath. She also complains of not being able to lift the same amount of weight. Which is the diagnosis?

a. hypothyroidism
b. hyperthyroidism
c. diabetes mellitus type II
d. scleroderma
e. rheumtoid arthritis
f. hypopituitarism
g. AS
h. degenerative joint disease

76. According to the previous question, which are two clinical findings for the diagnosis?

a. positive Allen's test
b. polydipsia
c. fever
d. Haygarth's nodes
e. gigantism
f. polyuria
g. positive Turyn's sign
h. silver-like flakes
i. dagger sign
j. Romanus lesion

77. According to question 75, which two tests would you give to this patient?

a. glucose tolerance test
b. Wright's test
c. urinalysis
d. Shilling test
e. VDRL
f. HCG
g. endoscopy
h. occult blood test
i. Milgrim's test
j. Beery's test

78. Charcot's triad is present in a 42-year-old female pastry chef. Scotoma,

muscle spasticity and weakness are present. Emotional outbursts are common. Which of the following is the most probable condition?

a. Parkinson's disease
b. rheumatoid arthritis
c. bipolar disorder
d. anxiety disorder
e. multiple sclerosis
f. Huntington's chorea
g. scleroderma
h. lymphosarcoma

79. Which of the following is not associated with multiple sclerosis?

a. demyelination of CNS
b. nystagmus
c. exacerbations and remissions
d. fatigue
e. intention tremor
f. paralysis
g. scanning speech
h. cogwheel rigidity

80. A 22-year-old female student has epilepsy. Pick two signs or symptoms of this condition.

a. Petit mal seizures
b. Grand mal seizures
c. fever
d. butterfly rash
e. dopamine deficiency
f. resting tremor
g. joint pain
h. Kayser Fleischer rings

81. Bowing deformity and rachitic rosary are present in a 2-year-old male. Paint brush metaphysis of the femur is also present. Which is the most likely diagnosis?

a. osteomalacia
b. AS
c. DJD
d. RA
e. scleroderma
f. rickets
g. hypothyroidism
h. hyperthyroidism

82. Based on the previous question, what are two clinical characteristics of this diagnosis?

a. pseudofracture
b. positive Sicard's sign
c. acromegaly
d. gigantism
e. increased alkaline phosphatase
f. punched out bone lesions
g. barrel chest
h. eosinophilic granuloma
i. fever
j. dagger sign

83. A 75-year-old female is suspected of having multiple myeloma. Which of the following lab tests are most appropriate for this condition?

a. ELISA
b. ANA
c. VDRL
d. reverse A/G ratio
e. Shilling test
f. Bence Jones protein urea
g. HCG
h. amylase
i. TSH
j. specific gravity

84. A 23-year-old male presents with pitted finger nails. Radiographic findings indicate cocktail sausage digit and pencil in cup deformity. In addition, ESR is elevated. Which is the most likely condition? Also, name another characteristic of this diagnosis.

a. AS
b. scleroderma
c. psoriatic arthritis
d. DJD
e. RA
f. Romanus lesion
g. calcinosis
h. conjunctivitis
i. dry, silver scales
j. geodes

85. A 32-year-old pregnant woman presents with pain and paresthesia on the anterior lateral portion of the right leg. Which of the following is the most probable condition?

a. multiple sclerosis
b. meralgia paresthetica
c. toxemia of pregnancy
d. syringomyelia
e. scleroderma
f. sarcoidosis
g. osteosarcoma
h. AS

86. According to the previous question, which are two other causes for the condition described?

a. demyelination of white matter
b. gliosis
c. poison
d. tight belts
e. obesity
f. lumbar disc herniation
g. lumbar posterior prolapse
h. fatigue
i. fever
j. infection

87. According to question 85, which nerve is primarily affected in this condition?

a. superior gluteal nerve
b. obturator nerve
c. inferior gluteal nerve
d. sciatic nerve
e. deep peroneal nerve
f. ulnar nerve
g. radial nerve
h. lateral femoral cutaneous nerve

88. A 43-year-old woman complains of a retracted right nipple. A hard, irregular, immoveable small tumor is found in the right upper quadrant of the right breast. Which is the most probable condition?

a. fibroadenoma
b. fibrocystic disease
c. cystic fibrosis
d. breast cancer
e. gynecomastia
f. galactorrhea
g. Paget disease
h. clinically insignificant

89. Hyperacusis and loss of sensation of sweet, sour and salt are assessed clinically in a 58-year-old retired postal worker. In addition, ptosis is present and drooping muscles on the right side of the face is present. Which is the diagnosis?

a. cerebral palsy
b. trigeminal neuralgia
c. Guillain-Barre syndrome
d. Bell's palsy
e. Parkinson's disease
f. multiple sclerosis
g. chronic fatigue syndrome
h. Paget's disease

90. Loss of pain and temperature in a shawl-like distribution over the torso is present in a 40-year-old male. The most likely condition is:

a. Shy-Drager syndrome
b. multiple myeloma
c. myasthenia gravis
d. neurotmesis
e. Guillain-Barre syndrome
f. syringomyelia
g. Charcot's triad
h. Barre-Lieou syndrome

91. A 25-year-old male complains of loss of sensation, tingling and weakness in each leg that started four days ago. Approximately, one month previous, this patient was treated for a small laceration in the right anterior thigh. Other clinical evidence characterizes this condition as an acute polyneuropathy. Which is the next step?

a. adjust C1
b. refer to massage therapist
c. adjust the sacrum
d. call 911
e. treat with ultrasound
f. treat with TENS
g. refer to psychotherapist
h. adjust extremities

92. According to the previous question, what is the diagnosis based on the description?

a. syringomyelia
b. myasthenia gravis
c. Guillain-Barre syndrome
d. multiple sclerosis
e. fibrous dysplasia
f. DJD
g. heart disease
h. sarcoidosis

93. Based on the previous question, which are two clinical manifestations of this diagnosis?

a. blood pressure fluctuation
b. weakness in muscles of respiration
c. migraine headaches
d. loss of pain and temperature in shawl-like distribution
e. geodes
f. osteophytes
g. angina
h. demyelination of white matter of CNS
i. emotional outbursts
j. muscle calcification

94. A 45-year-old male presents with headache, vertigo and tinnitus. In addition, George's test is positive. Previous tests reveal altered blood flow in the vertebral artery. Which is the most probable condition?

a. myasthenia gravis
b. Bell's palsy
c. cerebral palsy
d. multiple sclerosis
e. Barre-Lieou syndrome
f. diabetes mellitus type II
g. pancreatitis
h. amyotrophic lateral sclerosis

95. A 19-year-old male has amyotrophic lateral sclerosis. In which location would this motor neuron disease usually begin?

a. feet
b. chest
c. neck
d. lumbar spine
e. thigh
f. hands
g. skull
h. shoulders

96. An 18-year-old female presents with puffy eyelids, pain in the abdomen and shortness of breath. Lab tests reveal proteinuria. Which of the following is the most probable condition?

a. cerebral palsy
b. Barre-Lieou syndrome
c. myxedema
d. nephrotic syndrome
e. Charcot's triad
f. syphilis
g. amyotrophic lateral sclerosis
h. myasthenia gravis

97. Which are two findings associated with the previous condition described in question 96?

a. low blood albumin levels
b. Argyll-Robertson pupil
c. paralysis
d. irregular blood flow in vertebral artery
e. hyperreflexia in the legs
f. cranial nerve dysfunction
g. low urine levels of sodium
h. migraine headache
i. angina
j. gummas

98. A 48-year-old construction worker complains of ringing in the right ear and episodes of extreme vertigo. Ear infection is absent. Which is the most likely condition?

a. otosclerosis
b. chronic otitis media
c. acute otitis media
d. Meniere's disease
e. brain tumor
f. transient ischemic attacks
g. Reiter's disease
h. presbycusis

99. A 23-year-old male presents with conjunctivitis and urethritis. Lab tests reveal +HLA-B27. Which is the most probable condition? Also, name a another characteristic of this condition.

a. scleroderma
b. pelvic inflammatory disease
c. Reiter's syndrome
d. AS
e. hypothyroidism
f. Scarlet fever
g. arthritis
h. dry, brittle hair
i. Raynaud's phenomenon
j. dagger sign

100. A 33-year-old woman pregnant woman has edema and hypertension. Urinalysis reveals proteinuria. Which is the diagnosis?

a. reflex sympathetic dystrophy syndrome
b. pre-eclampsia
c. RA
d. malignancy
e. syphilis
f. pelvic inflammatory disease
g. rheumatic fever
h. SLE

101. A 32-year-old woman presents with xerophthalmia, xerostomia and arthritic changes in the hands. Rheumatoid factor is present. The most likely condition is which of the following?

a. Sjogren's syndrome
b. multiple sclerosis
c. degenerative joint disease
d. psoriasis
e. rubella
f. syphilis
g. stasis dermatitis
h. Felty's syndrome

102. Based on the previous question, which of the following is associated with the above diagnosis?

a. butterfly rash
b. photosensitivity
c. maculopapular rash
d. silver scales
e. sicca complex
f. fibromyalgia
g. varicocele
h. pre-eclampsia

103. A 33-year-old woman presents with enlarged axillary lymph nodes. She has a high fever for the past 4 days and her weight has decreased by over 10% in the last 2 months. In addition, she complains of night sweats. Which is the diagnosis?

a. Lyme disease
b. pancreatitis
c. peptic ulcer
d. Hodgkin's disease
e. mitral stenosis
f. hepatitis A
g. SLE
h. rheumatoid arthritis

104. According to the previous question, which of the following is associated with this condition?

a. Reed-Sternberg cells
b. candidiasis
c. rheumatic heart disease
d. hiatal hernia
e. LE cell
f. Coombs test
g. positive Libman's test
h. Addison's disease

105. A 52-year-old lawyer presents with buffalo hump, purple striae and weakness. Blood pressure is 175/98 mm Hg. Identify the most probable condition?

a. myxedema
b. Cushing's syndrome
c. cretinism
d. acromegaly
e. SLE
f. Lyme disease
g. Grave's disease
h. pancreatitis

106. Based on the previous question, this condition is most associated with which of the following?

a. Felty's syndrome
b. hypoadenalism
c. hyperthyroidism
d. lymphoma
e. AS
f. RA
g. hypercortisolism
h. multiple myeloma

107. A 32-year-old male driver was hit from behind by a large, box truck. This patient suffered moderate whiplash. Which of the following would most likely be positive?

a. Guilland's sign
b. Bikele's sign
c. Kernig sign
d. Forestier's Bowstring Sign
e. Hoover's sign
f. Rust sign
g. Ballottement test
h. Lachman test

108. A 45-year-old weightlifter is suspected of having an L4 disc herniation. Which of the following is most appropriate to test for this condition?

a. Kernig's
b. Cozen's
c. Mill's
d. McMurray's
e. Kemp's
f. Ortolani's
g. Dawbarn's
h. Magnuson's

109. A 23-year-old female presents with a bulls eye rash on the right anterior thigh. In addition, the patient has chills and fatigue. Identify the correct diagnosis.

a. herpes simplex
b. syphilis
c. gout
d. Lyme disease
e. sarcoidosis
f. rheumatic fever
g. pseudogout
h. psoriasis

110. According to the previous question, which are two possible complications of this condition?

a. meningitis
b. pericarditis
c. AS
d. RA
e. DJD
f. gummas
g. chancre
h. fluid-filled blisters
i. podagra
j. urate deposits

111. A 49-year-old carpenter has painless, knee joint destruction. Also,

there is debris and erosion in the knee joint. Which is the most likely condition?

a. CPPD
b. gout
c. neurotrophic arthritis
d. AS
e. sarcoidosis
f. fibrous dysplasia
g. sickle-cell anemia
h. osteoporosis

112. Hilar lymphadenopathy is present in a 41-year-old Scandinavian female. Several granulomas are found in the lungs. This patient complains of joint pain in the hands. Identify the correct diagnosis.

a. neurotrophic arthritis
b. RA
c. DJD
d. sarcoidosis
e. Osteitis Deformans
f. scleroderma
g. histiocytosis X
h. tuberculosis

113. According to the previous question, which is usually a beginning sign or symptom of this condition?

a. podagra
b. bone destruction
c. Haygarth's nodes
d. Raynaud's phenomenon
e. enlarged hat size
f. pannus formation
g. pneumothorax
h. erythema nodosa

114. A 65-year-old male presents with vertebral plana of T6 and several wedged vertebrae in the cervical spine. McConkey's sign is present. Which is the most likely diagnosis?

a. sickle-cell anemia
b. tuberculosis
c. osteoporosis
d. scleroderma
e. acromegaly
f. fibrous dysplasia
g. RA
h. AS

115. According to the previous question, name two other findings observed in this condition.

a. Romaus lesion
b. Pott's disease
c. Fish vertebrae
d. hyperkyphosis
e. increased jaw size
f. increased size of hands
g. severe chest pain
h. priapism
i. painful urination
j. dermatitis

116. A 72-year-old woman is suspected of having multiple myeloma. Which of the following is a finding consistent with this condition? (Pick two choices.)

a. Reverse A/G ratio
b. Positive Reed Sternberg cells
c. increased SGOT
d. Bence Jones proteinuria
e. increased HCG
f. decreased HCG
g. positive FTA
h. increased triiodothyronine

117. A 62-year-old male presents with deep, lumbar pain. In addition, the patient feels a pulsating sensation in the abdomen. Which of the following conditions should most likely be suspected?

a. lung carcinoma
b. pancreatitis
c. multiple myeloma
d. aortic aneurysm
e. diabetes mellitus type II
f. osteoporosis
g. sarcoidosis
h. tuberculosis

118. A 15-year-old male patient presents with abdominal tenderness over McBurney's point. In addition, he has episodes of vomiting over the past 6 hours. Urination is also quite painful. Which of the following is the most appropriate treatment protocol?

a. refer to hospital emergency room
b. adjust sacrum
c. adjust T6
d. adjust upper cervicals
e. massage abdomen
f. perform malingering tests
g. use Russian electrical stimulating currents
h. place hot packs over abdomen

119. According to the previous question, which is most probable diagnosis?

a. tuberculosis
b. leukemia
c. pancreatitis
d. aortic aneurysm
e. appendicitis
f. chondrosarcoma
g. osteosarcoma
h. cystic fibrosis

120. Osgood's Schlatter's disease is suspected in a 21-year-old college soccer player. Which part of the body is primarily affected?

a. skull
b. tibial tuberosity
c. calcaneus
d. ribs
e. femur
f. ilium
g. humerus
h. ischial tuberosity

121. A 15-year-old male presents with a hyperkyphosis. Upon x-ray evaluation, there is end-plate erosion coupled with Schmorl's nodes in the T6-T9 areas of the spine. There is anterior wedging between T4-T8. Which is the most probable condition?

a. Scheuermann's disease
b. AS
c. RA
d. DJD
e. Reiter's syndrome
f. muscular dystrophy
g. Legg- Calve-Perthes disease
h. hypothyroidism

122. A 42-year-old female librarian presents with pain and moderate paresthesia in shoulder and arm. Adson's test is positive. Which is the most probable diagnosis?

a. DJD
b. scleroderma
c. thoracic outlet syndrome
d. C6 herniated disc
e. abdominal aneurysm
f. RA
g. fibromyalgia
i. multiple sclerosis

123. According to the previous question, identify three tests that are associated with the diagnosis?

a. Halstead maneuver
b. Allen's test
c. Yergason's test
d. Hibb's test
e. Kemp's test
f. Reverse Bakody maneuver
g. Dugas' test
h. Codman's test
i. Bracelet test
j. Milgrim's test

124. According to question 122, what is one cause for this condition?

a. car accident
b. Raynaud's phenomenon
c. diet high in saturated fat
d. hypothyroidism
e. cervical rib
f. demyelination of CNS
g. overuse of joints
h. hyperthyroidism

125. An 8-year-old male is suspected of having bacterial meningitis. What are two signs or symptoms most closely related to this condition?

a. chancre
b. gummas
c. high fever
d. stiff neck
e. resting tremor
f. transient ischemic attack
g. murmurs
h. osteoma
i. acoustic neuroma
j. pityriasis rosea

126. Which three of the following would most likely be positive with bacterial meningitis?

a. Kernig sign
b. Brudzinski sign
c. Guilland's sign
d. sternal compression test
e. Lewin Supine test
f. Libman's test
g. Fouchet's sign
h. McMurray sign
i. Thomas test
j. Anvil test

127. A 52-year-old woman presents with hirsutism of the face and legs. Buffalo hump, moon face and striae on outer thigh are also apparent. Identify this condition.

a. diabetes mellitus type II
b. Tay Sachs disease
c. Addison's disease
d. fibromyalgia
e. diabetes insipidus
f. scleroderma
g. Cushing's syndrome
h. myxedema

128. According to the previous question, which are two other clinical manifestations of this condition?

a. telangiectasia
b. idiocy
c. more than 10 tender points on the body
d. Raynaud's phenomenon
e. weight gain
f. fever
g. stiff neck
h. bronze skin
i. positive Minor's sign
j. positive Rust's sign

129. A 42-year-old woman presents with dizziness, blue-black discoloration of the areola, and weight loss. In addition, the skin has a bronze-like color. Which of the following is the cause of this condition?

a. adrenal cortical insufficiency
b. stroke
c. transient ischemic attacks
d. drop attacks
e. diabetes mellitus
f. hypothyroidism
g. scleroderma
h. sickle-cell anemia

130. Name two other signs or symptoms related to the condition described in question 129?

a. osteoma
b. positive Minor's sign
c. low blood pressure
d. obesity
e. multiple myeloma
f. positive Roos test
g. biotin deficiency
h. vitamin C deficiency
i. weakness of muslces
j. selenium deficiency

131. According to the description in question 129, which is the most accurate diagnosis?

a. diabetes mellitus
b. hypothyroidism
c. hyperthyroidism
d. Addison's disease
e. diabetes insipidus
f. tuberculosis
g. acromegaly
h. Paget's disease

132. A 21-year-old female has production of thick, sticky mucous secretions leading to blockage of pancreatic ducts. The digestion of fats and proteins are severely slowed. Which is the most likely condition?

a. tuberculosis
b. cystic fibrosis
c. pulmonary embolism
d. diphtheria
e. respiratory distress syndrome
f. sarcoidosis
g. pleurisy
h. lymphosarcoma

133. Which portion of the population is most commonly affected with the condition described in the previous question? Also, name primarily which location of the body is affected by this condition.

a. White
b. African-American
c. Hispanic
d. American-Indian
e. Hawaiian
f. skull
g. femur
h. humerus
i. iris
j. exocrine glands

134. According to question 132, which organs of the body are primarily affected by this condition? (Identify three choices.)

a. heart
b. lungs
c. intestines
d. kidney
e. eyes
f. ears
g. pancreas
h. spinal cord

135. A 32-year-old patient who presently has pneumonia also has a stabbing pain in the chest. Upon auscultation, it seems the visceral and parietal pleural are rubbing together. The patient also has rapid, shallow breathing. Which is the most probable condition?

a. pancreatitis
b. RA
c. DJD
d. cystic fibrosis
e. pleurisy
f. lung cancer
g. osteosarcoma
h. lymphosarcoma

136. A 2-year-old male has the flu. However, the infant has developed a bark-like cough. Which is the diagnosis?

a. croup
b. meningitis
c. cystic fibrosis
d. leukemia
e. pleurisy
f. emphysema
g. cor pulmonale
h. emphyema

137. A 45-year-old truck drive involved in an automobile accident has a suspected L4 herniated disc. Based on the L5 neurological level, each of the three muscles tested have a muscle grade of 4. Which muscles have been tested for this L4 disc herniation? (Identify three choices.)

a. gluteus maximus
b. extensor digitorum longus
c. gluteus medius
d. peroneus longus
e. peroneus brevis
f. soleus
g. gastrocnemius
h. extensor digitorum brevis
i. trapezius
j. SCM

138. According to the above question based on the L5 neurological level, which reflex is most appropriate to be tested? Also, name a muscle action that is tested for the L5 neurological level.

a. patella
b. Achilles' tendon
c. none
d. biceps
e. triceps
f. forearm flexion
g. hip extension
h. leg abduction
i. foot plantar flexion
j. knee extension

139. Which are two muscle actions for the S1 neurological level?

a. foot plantar flexion
b. foot dorsiflexion
c. hip extension
d. leg abduction
e. finger extension
f. finger abduction
g. hip adduction
h. leg abduction
i. forearm extension
j. shoulder abduction

140. A 49-year-old weightlifter has a C5 herniated disc. Which muscle action should be tested?

a. wrist flexion
b. shoulder abduction
c. forearm flexion
d. finger flexion
e. finger adduction
f. finger abduction
g. wrist extension
h. finger extension

141. Pick three choices that should be tested for sensation at the C6 neurological level?

a. medial forearm
b. lateral forearm
c. lateral upper arm to elbow crease
d. posterior calf
e. index finger
f. fifth digit
g. lateral palm
h. lateral leg

142. Foot dorsiflexion is tested in a 24-year-old soccer player. This muscle action is given a grade of 2. Pick the correct description of grade 2. Also, what neurological level does this muscle action belong to?

a. normal response
b. contractility is absent
c. against gravity with moderate resistance, complete range of motion is observed
d. gravity is eliminated, complete range of motion is observed
e. complete range of motion is not observed
f. S1
g. S2
h. L2
i. L1
j. L4

143. Based on the previous question, what locations should be tested for sensation with a pinwheel? (Pick three choices.)

a. umbilicus
b. groin
c. nipple line
d. xiphoid process
e. anterior medial thigh
f. medial 1st toe
g. medial foot
h. medial knee to medial ankle

144. The corneal reflex is performed on a 41-year-old female patient. Which two cranial nerves are being tested?
a. I
b. II
c. III
d. IV
e. V
f. VI
g. VII
h. VII
i. VIII
j. X

145. A 34-year-old construction worker who is applying for worker's compensation is suspected of faking his low back injury. Which three of the following can be performed to rule out this suspected malingering patient?
a. Kernig's test
b. Bonnet's sign
c. Adson's test
d. Halstead's test
e. Schepelmann's sign
f. Trunk Rotational test
g. Burns Bench test
h. Wright test
i. Mankopf's maneuver
j. Trendelenburg test

146. A Babinski response is apparent in a 29-year-old male when stroking the lateral malleolus with the handle of a reflex hammer. Which pathological reflex does this best represent?
a. plantar
b. Tromner's
c. Gordon's
d. Chaddock's
e. Schaffer's
f. Hoffman's
g. Snout
h. Ankle clonus

147. Upon a stroking the plantar surface of the foot, dorsiflexion of the 1st toe and fanning of the other toes is apparent. Which pathological reflex does this most accurately describe?
a. Abdominal
b. Babinski's
c. Chaddock's
d. Ankle clonus
e. Hoffman's
f. Oppenheim's
g. Schaffer's
h. Rossolimo's

148. A 34-year-old woman complains of pain and tingling in the right arm and shoulder that wakes her during sleep. Occasionally, using a can opener is quite painful. Work history reveals extensive use of the computer for a period of 10 years. Which of the following will most likely be positive in this patient? Choose two choices.

a. Tinel's sign
b. Phalen's sign
c. Mill's test
d. Kaplan's sign
e. Finsterer's sign
f. Dawbarn sign
g. Cozen's test
h. Roos test
i. Adson's test
j. Dugas' test

149. According to the previous question, which is the most probable diagnosis?

a. thoracic outlet syndrome
b. RA
c. tennis elbow
d. golfer's elbow
e. subacromial bursitis
f. carpal tunnel syndrome
g. Kienbock's disease
h. malingering

150. According to the previous question, which is the most likely cause of this condition?

a. cervical rib
b. medial epicondylitis
c. lateral epicondylitis
d. aseptic necrosis of the lunate
e. compression of the median nerve
f. acute sports injury
g. food allergy
h. fracture

151. Which of the following tests for cervical ligamentous sprain?

a. Naffziger's test
b. O'Donoghue maneuver
c. Allen's test
d. Lhermitte's sign
e. Bakody sign
f. Codman's sign
g. Dugas' test
h. Roos test

152. A 51-year-old female accountant presents with acid reflux. She complains of heartburn that gets worse when lying down. The stomach is suspected of passing through the esophageal hiatus. Which is the diagnosis?

a. Chron's disease
b. duodenal ulcer
c. cholecystitis
d. cholelithiasis
e. chronic pancreatitis
f. Hodgkin's disease
g. hiatal hernia
h. glomerulonephritis

153. Based on the previous question, which two of the following is this condition associated with?

a. pregnancy
b. obesity
c. migraines
d. joint pain
e. Raynaud's phenomenon
f. MI
g. arrhythmias
h. mitral valve stenosis
i. syncope
j. fever

154. A 32-year-old female complains of frequent, bloody, watery diarrhea. Colonoscopy revealed an inflammation and areas of ulceration in the large intestine. However, no cancer is present. Which is the most likely condition? Also, identify another sign or symptom of this condition.

a. diverticulitis
b. hiatal hernia
c. ulcerative colitis
d. hepatitis A
e. cholelithiasis
f. abdominal cramps
g. thoracic outlet syndrome
h. hematoma
i. transient ischemic attacks
j. osteoma

155. Identify two ranges of motion of the shoulder.

a. supination
b. dorsiflexion
c. pronation
d. eversion
e. inversion
f. abduction
g. adduction
h. plantar flexion
i. radial deviation
j. ulnar deviation

156. Which of the following does not have a range of motion of extension?

a. shoulder
b. wrist
c. cervical spine
d. lumbar spine
e. hip
f. knee
g. elbow
h. ankle

157. The cervical sympathetics T1-T3 need to be tested. Therefore, the patient's posterior neck is pinched and the eyes dilate. Which reflex does

this best represent?

a. plantar
b. ciliospinal
c. carotid
d. uvular
e. corneal
f. Geigel's
g. jaw jerk
h. pectoralis

158. A 32-year-old male presents with a herniated L3 disc. The patella reflex is given a grade of zero on the Wexler scale. Which sign does this best represent?

a. Chaddock's wrist sign
b. Gordon's finger sign
c. Codman's sign
d. Kaplan's sign
e. Dawbarn's sign
f. Finsterer's sign
g. Phalen's sign
h. Westphal's sign

159. Which two lung conditions will usually present with a percussive note of hyperresonant?

a. emphysema
b. bronchitis
c. pneumonia
d. pleurisy
e. pnuemothorax
f. atelectasis
g. bronchiectasis
h. hemothorax

160. A 61-year-old male is diagnosed with cor pulmonale. Which are two signs or symptoms of this condition?

a. pulmonary emphysema
b. hypertrophy of right ventricle
c. croup
d. cystic fibrosis
e. fever
f. epiglottitis
g. bronchiolitis
h. spoon nails
i. pustules
j. Rust's sign

161. A 49-year-old male is suspected of having kidney stones. Which of the following might have confirmed these findings?

a. rebound tenderness
b. Murphy's sign
c. fluid wave
d. Obturator's sign
e. Murphy's punch
f. venous hum
g. Rovsing sign
h. psoas sign

162. A 23-year-old male has appendicitis. Which three of the following will be positive?

a. Murphy's sign
b. Murphy's punch
c. venous hum
d. psoas sign
e. Rovsing sign
f. fluid wave
g. Bakody sign
h. McMurray sign
i. obturator's sign
j. Kaplan's sign

163. A 22-year-old female personal trainer clinically has a mid-systolic click upon auscultation. Her history reveals bouts with rheumatic fever. Which is the most likely diagnosis?

a. myocardial infarction
b. pulmonary edema
c. angina pectoris
d. cirrhosis
e. pleurisy
f. mitral valve prolapse
g. hyperglycemia
h. ascites

164. A 71-year-old retired, female postal worker presents with festinating gait, pill rolling tremors and drooling. Which of the following is the most probable condition?

a. Parkinson's disease
b. multiple sclerosis
c. amyotrophic lateral sclerosis
d. myasthenia gravis
e. syringomyelia
f. transient ischemic attack
g. stroke
h. cerebral palsy

165. Identify two other signs or symptoms associated with this condition.

a. high, monotone voice
b. scotomas
c. intention tremor
d. drop attacks
e. Charcot's joint
f. pigeon chest deformity
g. mask-like stare
h. Horner's syndrome
i. weakened muscles innervated by cranial nerves
j. scissors gait

166. A 59-year-old bartender presents with persistent cough coupled with blood-streaked sputum. Which two of the following should be performed next?

a. cervical adjustment
b. chest x-ray
c. sputum cytology
d. spinal instrumentation
e. Jolly test
f. therapeutic ultrasound
g. nutritional therapy
h. spinal traction

167. Based on the question 165, what is most likely diagnosis?

a. emphysema
b. syringomyelia
c. bronchogenic carcinoma
d. hematoma
e. histiocytosis X
f. multiple myeloma
g. bronchitis
h. croup

168. A 34-year-old male has recently been involved in a boating accident. Since there is no hospital within miles, he decides to seek your professional guidance. He presents with bilateral sciatic pain, loss of bladder and bowel control, and saddle paresthesia. Which of the following is the most likely diagnosis?

a. Barre-Lieou syndrome
b. amyotrophic lateral sclerosis
c. brain tumor
d. cauda equina syndrome
e. multiple sclerosis
f. cystic fibrosis
g. AS
h. RA

169. A 42-year-old male presents with tinnitus, blurred vision, vertigo and dysphagia. George's test is positive. Which of the following would you perform to further confirm your diagnosis? (Choose two choices.)

a. Hautant's test
b. Roos test
c. Burn's Bench test
d. Ballottement test
e. Noble compression test
f. Thomas test
g. Thompson's test
h. Dekleyn's test
i. Mill's maneuver
j. Apprehension test

170. According to the description in the previous question, which is the diagnosis?

a. Charcot's triad
b. cauda equina syndrome
c. multiple sclerosis
d. thoracic outlet syndrome
e. amyotrophic lateral sclerosis
f. Shy-Drager syndrome
g. Barre-Lieou syndrome
h. Wallenberg's syndrome

171. Which is one cause of the condition described in question 169?

a. stenosis of vertebral artery
b. intention tremor
c. demyelination of CNS
d. migraine
e. upper and motor neuron disease
f. myoclonus
g. Erb's paralysis
h. claw hand

172. A 34-year-old marathon runner presents with crepitus of the knee joint, articular cartilage damage and pain. Which is the most likely diagnosis?

a. adhesive capsulitis
b. chondromalacia patallae
c. RA
d. AS
e. osteoma
f. hematoma
g. multiple sclerosis
h. tarsal tunnel syndrome

173. Which of the following should be done specifically for the previous condition described?

a. strengthen shoulders
b. strengthen SCM
c. recommend cervical pillow
d. strengthen vastus medialis
e. wear brace
f. treat as medical emergency
g. complete bed rest for two months
h. Milwaukee brace

174. A 51-year-old woman is diagnosed with hyperparathyroidism. Identify two findings associated with this condition.

a. dagger sign
b. increased energy level
c. salt and pepper skull
d. rugger jersey spine
e. hematoma
f. stenosis of carotid artery
g. meningeal irritation
h. osteoma
i. multiple myeloma
j. resting tremor

175. A 45-year-old male office worker presents with suspected sciatica. Which three of the following will further confirm this suspected diagnosis?

a. Ely's sign
b. Buckling sign
c. Braggard's test
d. Well Leg raiser
e. Heel walk
f. Hibb's test
g. Patrick test
h. Sicard's test
i. Kemp's test
j. Dekleyn's test

176. A 23-year-old male has an L4 disc protrusion which causes pain and paresthesia down the right leg. Which two of the following should be positive?

a. Well Leg raiser
b. Fajersztajn's test
c. Patrick test
d. Hibb's test
e. Halstead's test
f. Eden's test
g. Wright test
h. Allen's test
i. Adam's sign
j. Bracelet test

177. A 34-year-old male has suspected *hepatitis A* after eating at a fast food restaurant. Which three lab tests are most appropriate for this patient?

a. Mantoux test
b. HLA-B27
c. LDH
d. CPK
e. amylase
f. SGPT
g. Paul-Bunnell
h. SGOT
i. bilirubin
j. ASO titer

178. A 36-year-old male presents with jaundice. He complains of having clay colored stools and brownish color to his urine. Alkaline aminotransferase is elevated. Which of the following is the best diagnosis?

a. pernicious anemia
b. subacute bacterial endocarditis
c. glomerulonephritis
d. diabetes mellitus
e. leukemia
f. hepatitis B
g. pancreatitis
h. prostate cancer

179. According to the diagnosis in the previous question, which is the mode of transmission or cause of the disease?

a. hereditary
b. eating too many simple sugars over time
c. meat based diet
d. sexual intercourse/IV drug users
e. history of mumps
f. alcoholism
g. kidney degeneration
h. staphylococcal infection

180. What are two other conditions the diagnosis in question 178 can lead to?

a. cirrhosis
b. rheumatoid arthritis
c. psoriasis
d. croup
e. amyotrophic lateral sclerosis
f. liver cancer
g. C.P.P.D.
h. cellulitis
i. broncholiths
j. hordeolum

181. A 43-year-old teacher with severe gingivitis complains of rapid heartbeat, tiredness and weight loss for the past 2 months. Medical history reveals anemia. Which is the most probable diagnosis?

a. subacute bacterial endocarditis
b. prostate cancer
c. osteoarthritis
d. pleurisy
e. croup
f. multiple myeloma
g. emphysema
h. nephrotic syndrome

182. Which is the usual cause of the above description described?
a. second-hand smoke
b. joint overuse
c. meat based diet
d. damaged heart valves
e. Bence-Jones proteinuria
f. reverse A/G ratio
g. obesity
h. aneurysm

183. A 29-year-old male presents complains of tender knee and wrist joints. Urethritis and conjuctivitis are also present. Which is the most relevant question to ask this patient based on the description?
a. Do you eat red meat?
b. Do you drink red wine?
c. Did you ever have rheumatic fever?
d. Did you or your sexual partners ever have venereal disease?
e. Do you have heart disease?

184. Based on the previous question, what is the diagnosis?
a. Wegener's granulomatosis
b. temporal arthritis
c. gout
d. pseudogout
e. Reiter's syndrome
f. ankylosing spondylitis
g. RA
h. DJD

185. What is the cause for the diagnosis in the previous question?
a. chlamydial infection
b. too much red meat
c. rheumatic fever
d. joint overuse
e. aneurysm
f. temporal arthritis
g. anemia
h. pancreatitis

186. After respiratory excursion was declared clinically insignificant in a 63-year-old female student, tactile fremitus was performed. Tactile fremitus was found to be decreased. Which are two possible causes for this?

a. emphysema
b. stroke
c. bronchial obstruction
d. pancreatitis
e. cervical cancer
f. compressed lung
g. hiatal hernia
h. ankylosing spondylitis
i. kidney stones
j. peptic ulcer

187. Which three of the following would be positive in a patient with whiplash? This patient has increased pain with sneezing and defecating.

a. Braggard's
b. Naffziger
c. Swallowing
d. Valsalva
e. Drawer
f. Halstead's
g. Hibb's
h. Minor's
i. Libman's
j. Ely's

188. Dullness is heard over the affected lung fields of a 44-year-old male patient. Which are two possible causes of this?

a. atelectasis
b. emphymsema
c. cirrhosis
d. pleural effusion
e. nasal polyps
f. arcus senilis
g. cephalhematoma
h. cholecystitis
i. cholelithiasis
j. episcleritis

189. A 72-year-old male has limited diaphragmatic excursion. History reveals of being an ex-smoker for 30 years. In addition, no previous trauma or accidents have been reported by the patient Which is the most likely condition?

a. atelectasis
b. cirrhosis
c. pneumothorax
d. pneumonia
e. croup
f. cystic fibrosis
g. tracheomalacia
h. emphysema

190. What are two other causes of limited diaphragmatic excursion?

a. arcus senilis
b. diverticulitis
c. fractured rib
d. irritable bowel syndrome
e. rubella
f. chondrocalcinosis
g. abdominal ascites
h. thrombophlebitis
i. hepatitis A
j. gout

191. A 23-year-old woman presents with tenderness and pain in the right lower quadrant of the abdomen. She complains of having diarrhea, loss of weight and cramps in the abdomen. In addition, bilateral wrist joint inflammation and episcleritis are evident. Her medical history reveals anal fistulas. Lab values reveal elevated white blood cells. Which is the diagnosis?

a. irritable bowel syndrome
b. colon cancer
c. Crohn's disease
d. RA
e. scleroderma
f. pancreatitis
g. syphilis
h. gonorrhea

192. According to the previous diagnosis, what two portions of the body are most commonly affected?

a. ileum of small intestine
b. tongue
c. stomach
d. vagina
e. epiglottis
f. mouth
g. teeth
h. trachea
i. large intestine
j. esophagus

193. According to the question 191, what are two other physical signs of this diagnosis?

a. blebs
b. aphthous stomatitis
c. gallstones
d. erythema nodosum
e. bronchitis
f. gummas
g. shingles
h. hepatitis
i. Raynaud's phenomenon
j. polycythemia vera

194. A 32-year-old woman complains of a greenish-yellow pus flow from the vagina coupled with pain. In addition, she has nausea and bouts of vomiting for the past couple of days. Which is the most likely condition?

a. shingles
b. genital herpes
c. neurosyphilis
d. primary syphilis
e. gonorrhea
f. genital warts
g. genital candidiasis
h. meningitis

195. Based on the above diagnosis, which is the causative agent?

a. Herpes simplex I
b. Trichomonas vaginalis
c. Neisseria gonorrhoeae
d. Treponema pallidum
e. Candida albicans
f. Condylomata acuminata
g. Herpes simplex II
h. cytomegalovirus

196. Based on question 194, what are two other physical signs of this disease?

a. low blood sugar
b. jaundice
c. cyanosis
d. peritonitis
e. gummas on the leg
f. polyarthritis
g. painless sore around the vagina
h. fluid-filled blisters in genital area
i. cottage cheese-like vaginal discharge
j. thrush

197. A 23-year-old male notices a painless sore on his penis. In addition, there are enlarged inferior superficial inguinal lymph nodes, which are painless. Which is the most likely diagnosis?

a. genital warts
b. genital candidiasis
c. thrush
d. gonorrhea
e. tertiary syphilis
f. trichomoniasis
g. primary syphilis
h. herpes simplex II

198. Based on the previous description, which is the causative agent?

a. Trichomonas vaginalis
b. Neisseria gonorrhoeae
c. Treponema pallidum
d. Herpes simplex type I
e. Candida albicans
f. Granuloma inguinale
g. Cytomegalovirus
h. Condylomata acuminata

199. Based on question 197, what is another sign or symptom of this disease?

a. cottage cheese-like vaginal discharge
b. thrush
c. greenish-yellow frothy discharge
d. low blood sugar
e. tabes dorsalis
f. Kaposi's sarcoma
g. pediculosis pubis
h. hepatitis

200. A 2-year-old infant presents with cherry red spots in the retina, inattentiveness, and positive Babinski reflex. Which of the following is the

most probable condition?

a. multiple myeloma
b. phenykenonuria
c. sickle-cell anemia
d. Tay Sachs disease
e. Addison's disease
f. rickets
g. scurvy
h. TB

201. According to the previous question, identify two other signs or symptoms of this condition?

a. blindness
b. mental retardation
c. punched-out bone lesions
d. goiter
e. vitamin C deficiency
f. vitamin D deficiency
g. blood-streaked sputum
h. Pott's disease
i. butterfly rash
j. dagger sign

202. According to the question 200, which is the mechanism of injury in this condition?

a. lack of vitamin D
b. lack of vitamin C
c. hypothyroidism
d. hyperthyroidism
e. diabetic retinopathy
f. abnormal neuro-lipid metabolism
g. Mycobacterium tuberculosis
h. sickle-shaped red blood cells

203. A 45-year-old male postal worker presents with blood sugar levels of 198 mg/dl. His chief complaint is excessive urination. In addition he is moderately obese. Which are two other physical signs of this condition?

a. polyphagia
b. fever
c. torticollis
d. tic douloureux
e. variocele
f. wheals
g. ulcerative colitis
h. polydipsia
i. saber shin
j. neuroblastoma

204. Based on the previous question, which is the diagnosis?

a. hypothyroidism
b. AS
c. RA
d. DJD
e. diabetes mellitus type II
f. multiple sclerosis
g. heart disease
h. chronic lymphocytic leukemia

205. A 32-year-old woman has been diagnosed with Wilson's disease. Which is a cause of this condition? Also, name two physical signs of this

condition.

a. excessive magnesium
b. selenium excess
c. copper excess
d. zinc excess
e. iodine deficiency
f. hirsutism
g. Kayser-Fleischer ring
h. drooling
i. cirrhosis
j. erythema chronicum migrans

206. A 34-year-old female school teacher has contracted Lyme disease. Which are two physical signs of this condition?

a. oral thrush
b. peptic ulcer
c. erythma chronicum migrans
d. myalgia
e. butterfly rash
f. hepatitis A
g. Kayser-Fleischer ring
h. resting tremor
i. Argyll Robertson pupil
j. pencil-in-cup deformity

207. Upon auscultation of the heart, an opening snap is heard at the beginning of diastole. In addition, other clinical findings reveal an enlarged pulmonary artery, pulmonary hypertension, and tachycardia. Which of the following is the most probable diagnosis?

a. mononucleosis
b. croup
c. Hodgkin's disease
d. peptic ulcer
e. mitral stenosis
f. nephrotic syndrome
g. Wilson's disease
h. bronchitis

208. Based on the question 207, which is the most common cause of this condition?

a. smoking
b. excessive alcohol
c. copper excess
d. iron excess
e. diabetes mellitus
f. rheumatic fever
g. Epstein-Barr virus
h. pancreatitis

209. Identify two signs or symptoms observed in the diagnosis in question 207.

a. Raynaud's phenomenon
b. gallstones
c. hiatal hernia
d. Kayser-Fleischer ring
e. paroxysmal nocturnal dyspnea
f. scleroderma
g. bark-like cough
h. orthopnea

210. A 32-year-old female manager complains of multiple tender points in the muscles and joints of the extremities. She complains of not being able to go into a deep sleep for the last few months. She is very fatigued and anxious by her suspected upcoming promotion. Which of the following is the diagnosis?

a. chronic fatigue syndrome
b. fibromyalgia
c. RA
d. DJD
e. scleroderma
f. Reiter's syndrome
g. SLE
h. AS

211. Which of the following is absolutely not an appropriate treatment protocol for the above diagnosis?

a. ROM exercises
b. stretching
c. massage
d. warm whirlpool
e. thermotherapy
f. rest
g. cryotherapy
h. psychotherapy

212. Based on the description in question 210, what are two other signs or symptoms of this condition?

a. headache
b. muscle spasm
c. Scarlet fever
d. Klumpke's paralysis
e. Koplik spots
f. macule
g. melena
h. papilledema
i. Murphy's sign
j. osteomalacia

213. A 45-year-old female with varicose veins presents with dark brown skin in both ankles. Which of the following is the most probable condition?

a. lymphedema
b. thrombophlebitis
c. stasis dermatitis
d. lymphangitis
e. polycythemia vera
f. Reye's syndrome
g. rubeola
h. Scarlet fever

214. Based on the description in question 213, what are two other physical characteristics of this condition?

a. edema
b. proteinuria
c. pre-eclampsia
d. coryza
e. skin ulceration
f. projectile vomiting
g. encephalopathy
h. strawberry tongue

215. A 23-year-old woman presents with strawberry tongue, pharyngitis and pink-red rash on the abdomen. Which is the most probable condition?

a. hypothyroidism
b. Scarlet fever
c. fibromyalgia
d. Addison's disease
e. hyperthyroidism
f. Reye's syndrome
g. polycythemia vera
h. psoriasis

216. A 4-year-old male presents with Koplik spots and a deep red maculo-papular rash on the face, neck and chest. Which is the diagnosis?

a. rubella
b. Addison's disease
c. Scarlet fever
d. Reye's syndrome
e. croup
f. SLE
g. rubeola
h. shingles

217. According to the previous question, which are two other physical characteristics of this condition?

a. psoriasis
b. polycythemia vera
c. hypothyroidism
d. migraine
e. fever
f. butterfly rash
g. adrenal cortical insufficiency
h. cough
i. insulin shock
j. papilledema

218. A 45-year-old woman has chronic muscle degeneration with painful inflammation in the shoulders and hips. Raynaud's phenomenon is present. Lab values reveal increased SGOT and creatinuria. Which is the diagnosis?

a. systemic lupus erythematosus
b. polymyositis
c. fibromyalgia
d. polyartheritis
e. thrombophlebitis
f. Scarlet fever
g. scleroderma
h. chronic fatigue syndrome

219. Identify three other signs or symptoms of the condition described in the previous question.

a. heliotrope rash
b. calcinosis
c. sclerodactyly
d. fever
e. glaucoma
f. hot veins in the leg
g. dysphagia
h. Blumberg's sign
i. hydrocele
j. asthma

220. A 44-year-old woman presents with spider veins on the legs, sclerodactyly, Raynaud's phenomenon, and esophageal dysfunction. Which is the most probable condition?

a. psoriasis
b. Sjogren's syndrome
c. SLE
d. RA
e. DJD
f. fibromyalgia
g. AS
h. scleroderma

221. Name another sign observed in the condition described in question 220?

a. migraine
b. brain tumor
c. dagger sign
d. silver, dry scales
e. calcinosis
f. butterfly rash
g. gout
h. resting tremor

222. A 22-year-old female presents with photosensitivity, butterfly rash, and has inflammation of the joints of the hands similar to rheumatoid arthritis. Which of the following is the most likely diagnosis?

a. psoriatic arthritis
b. SLE
c. sclerodactyly
d. Reiter's syndrome
e. Still's disease
f. stasis dermatitis
g. fibromyalgia
h. DJD

223. Which of the following will help confirm the diagnosis in question 222? (Pick two choices.)

a. antinuclear antibodies
b. LE cell
c. SGOT
d. Downey cell
e. Bence-Jones proteinuria
f. amylase
g. lipase
h. triglycerides

224. According to question 222, what are two other signs or symptoms of this condition?

a. mouth sores
b. glaucoma
c. thrush
d. hiatal hernia
e. tremors
f. anemia
g. thrombophlebitis
h. pyloric stenosis
i. Rovsing's sign
j. amnesia

225. A 23-year-old college baseball outfielder has collided with the outfield wall trying to track down a fly ball. He has a suspected dislocated shoulder. Which of the following is most appropriate to test for this diagnosis? (Pick two choices.)

a. Roos test
b. Halstead maneuver
c. Apprehension test
d. Dugas test
e. Mill's test
f. Shrivel test
g. Finkelstein's test
h. Lewin Standing test
i. Beery's test
j. Lewin Punch test

226. A 22-year-old football player has a torn meniscus. Which of the following will most likely be positive?

a. Lachman test
b. Beery's test
c. Lewin Punch test
d. Shrivel test
e. Apley's compression test
f. Thompson test
g. Belt test
h. Nachlas test

227. A 22-year-old swimmer has a possible contracture of the tensa fascia lata. Which of the following is the most appropriate to test for this condition?

a. Faber-Patrick test
b. Trendelenberg test
c. Allis test
d. Ober test
e. Hibb's test
f. McMurray test
g. Bounce Home test
h. Thompson test

228. A 21-year-old college baseball player has medial epicondylitis. Which of the following will be positive?

a. Golfer's Elbow test
b. Cozen's test
c. Bounce Home test
d. Dugas test
e. Lachman's test
f. Beery's test
g. Thompson test
h. Lewin Punch test

229. A 48-year-old plumber presents with sciatica. Which two of the following will be positive?

a. Braggard's test
b. Fajerstajn's test
c. Hautant's test
d. Milgram's test
e. Straight leg raiser
f. Kemp's test
g. Dugas test
h. Yergason test

230. A 51-year-old male has a saddle thrombus of the aorta. He also complains of impotence. In addition, intermittent claudication and weak pulses are also present. Which is the diagnosis?

a. dissecting aneurysm
b. MI
c. appendicitis
d. diverticulitis
e. emphysema
f. bronchitis
g. Leriche's syndrome
h. angina pectoris

231. A 54-year-old housewife has a positive Murphy's punch. Which of the following does this indicate?

a. cholecystitis
b. MI
c. bronchitis
d. appendicitis
e. hernia
f. kidney stones
g. cirrhosis
h. gallstones

232. Which two conditions are associated with Cullen's sign?

a. appendicitis
b. kidney stone
c. cholecystitis
d. pancreatitis
e. MI
f. hemoperitoneum
g. intussusception
h. bronchitis
i. angina pectoris
j. emphysema

233. Which of the following best describes Cullen's sign?

a. rebound tenderness
b. ecchymosis around umbilicus
c. absent bowel sounds
d. venous hum
e. ecchymosis near kidneys
f. kidney stones
g. skull fracture
h. pitting edema

234. Which of the following signs is associated with cholecystitis?

a. Aaron
b. McBurney
c. Rovsing
d. Blumberg
e. Grey Turner
f. Cullen
g. Murphy
h. Kehr

235. A 72-year-old male presents with clubbing of the fingernails. Which is the most probable condition?

a. eczema
b. psoriasis
c. trichinosis
d. bronchogenic carcinoma
e. brain tumor
f. herpes simplex
g. anemia
h. osteoporosis

236. A 45-year-old male complains of difficulty swallowing. Which of the following cranial nerves is most appropriate to test?

a. optic
b. trigeminal
c. vagus
d. abducens
e. facial
f. spinal accessory
g. olfactory
h. occulomotor

237. A 33-year-old secretary has a muscle grade of 2 when testing the wrist extensors. Which neurological level does this correspond to?

a. C1
b C2
c. C4
d. C5
e. C6
f. C7
g. C8
h. T1

238. A 34-year-old fireman presents with a gnawing type pain in the epigastric area of the abdomen. Eating food does seem to relieve these symptoms. He reports that he smokes approximately one pack of cigarettes per day. In addition, he has blood type O. Which of the following would you recommend?

a. doppler
b. Barium contrast x-ray
c. lateral cervical view
d. Mantoux test
e. spirometer
f. biopsy
g. NCV
h. EMG

239. According to the description in the previous question, which is the most probable diagnosis?

a. gastric ulcer
b. stomach cancer
c. ulcerative colitis
d. appendicitis
e. duodenal ulcer
f. Peutz-Jegher's syndrome
g. diverticula
h. Gardener's syndrome

240. Based on the description in question 238, which is the closest differential diagnosis to this condition?

a. stomach cancer
b. diverticulitis
c. volvulus
d. gastric ulcer
e. appendicitis
f. Crohn's disease
g. cholecystitis
h. duodenal ulcer

241. Based on question 239, which is the mechanism of injury? Also, name the type of bacteria commonly found with this condition.
a. ulcer in duodenum
b. ulcer in lining of stomach
c. malignant stomach tumor
d. gallstones
e. twisting of loops of large intestine
f. staphylococcus aureus
g. helicobacter pylori
h. Listereria monocytogenes
i. hemophilus influenzae
j. Pseudomonas

242. Which of the following are most appropriate to perform in a patient with sacroiliac lesion?
a. Iliac compression test
b. Belt test
c. Shrivel test
d. Phalen's test
e. Mill's test
f. Cozen's test
g. Roos test
h. Dugas test
i. Yergason's test
j. Gaenslen's test

243. A 45-year-old male patient complains of low back pain. The following tests are positive: Lindner's, Bowstring sign, Sicard's and Turyn's. Which of the following is the most probable condition?
a. Kienbock's disease
b. sciatica
c. disc lesion
d. meniscus tear
e. inflammation of the pleura
f. weak gluteus medius muscle
g. thrombophlebitis
h. scoliosis

244. A 49-year-old female college professor complains of pain in the upper epigastric area after eating. This patient is blood type A. She also says she takes aspirin on a daily basis for the past few months to prevent heart disease. Which of the of the following is the most probable diagnosis?
a. hiatal hernia
b. gastric carcinoma
c. duodenal ulcer
d. pyloric stenosis
e. intussusception
f. Mallory Weiss syndrome
g. gastric ulcer
h. appendicitis

245. According to the previous question, pick two procedures that will confirm this diagnosis?

a. EMG
b. Mantoux test
c. NCV
d. endoscopy
e. EEG
f. biopsy
g. spirometer
h. barium contrast x-ray
i. aspiration
j. arterial blood gas

246. According to question 244, which is the usual location of injury? Also, name the type of bacteria found in this condition.

a. lesser curvature of stomach
b. upper curvature of stomach
c. duodenum
d. kidney stones
e. gallstones
f. helicobacter pylori
g. Hemophilus
h. Pseudomonas
i. E. Coli
j. Listeria monocytogenes

247. A 6-year-old female has bouts with projectile vomiting. There is pain after eating. Which is the most probable condition?

a. hiatal hernia
b. pyloric stenosis
c. stomach cancer
d. intussusception
e. leukemia
f. divertculitis
g. Peutz-Jegher's syndrome
h. hepatitis A

248. A 52-year-old woman complains of weight loss and blood-streaked stools. In addition, she has bouts of constipation and diarrhea and a history of ulcerative colitis. Which of the following is the most probable diagnosis?

a. Hirschprung's disease
b. hepatitis A
c. pancreatitis
d. cirrhosis
e. carcinoma of colon
f. Gardener's syndrome
g. Peutz-Jegher's syndrome
h. cholelithiasis

249. According to the description in the previous question, which three of the following will further confirm your diagnosis?

a. spirometer
b. NCV
c. cholecystography
d. barium contrast x-ray
e. aspiration
f. myelography
g. intravenous urography
h. colposcopy
i. colonoscopy
j. biopsy

250. A 12-year-old Caucasian male presents with brown lips and gums. In addition, colonoscopy reveals benign polyps in the stomach, large intestine and small intestine. Which is the condition?

a. Gardener's syndrome
b. ulcerative colitis
c. diverticula
d. Peutz-Jegher's syndrome
e. volvulus
f. Hirschprung's disease
g. hepatitis A
h. acute pancreatitis

251. A 42-year-old female chef complains of constant pain that radiates to the lower back. In addition, it gets worse while laying down. She has also been vomiting occasionally for the past couple of days. Cullen's sign is present. Which is the most likely condition?

a. acute pancreatitis
b. Gardener's syndrome
c. Hirschprung's disease
d. ulcerative colitis
e. hepatitis B
f. Peutz-Jegher's syndrome
g. volvulus
h. hiatal hernia

252. Based on the previous question, what are two other physical characteristics of this condition?

a. glomerulonephritis
b. pleurisy
c. diarrhea
d. headache
e. scotoma
f. volvulus
g. cervical cancer
h. tachycardia
i. irritable bowel syndrome
j. Gray-Turner's sign

253. Based on question 251, which are two findings of this condition?

a. Psoas sign
b. Murphy's sign
c. hyperglycemia
d. elevated serum amylase
e. diabetes mellitus
f. absent jaw jerk reflex
g. pigeon breast deformity
h. nystagmus
i. dysarthria
j. Bikele's sign

254. According to question 251, what are two common causes of this condition?

a. gallstones
b. alcoholism
c. migraines
d. cervical cancer
e. gingivitis
f. heart disease
g. brain tumor
h. sexual contact with infected partner
i. sacculation of bowel walls
j. sebaceous cyst

255. A 6-month-old male has decreased peristalsis, inflammation and sacculation of the bowel walls. Vomiting and constipation are evident. Which is the most likely diagnosis?

a. Gardener's syndrome
b. hepatitis B
c. cirrhosis
d. Hirschprung's disease
e. cholelithiasis
f. meningitis
g. irritable bowel syndrome
h. pancreatitis

256. Based on the previous question, which is a cause of this condition?

a. lack of ganglion cells in portion of large intestine
b. cirrhosis
c. gallstones
d. kidney stones
e. poor diet
f. ulcerative colitis
g. colon cancer
h. diverticulitis

257. A 32-year-old male presents with osteoma of the frontal sinus. Colonoscopy reveals polyps in the large intestine. In addition, fibromas are also apparent. Which of the following is the diagnosis?

a. ulcerative colitis
b. gastric ulcer
c. colon cancer
d. hepatitis C
e. cholelithiasis
f. cirrhosis
g. Gardener's syndrome
h. irritable bowel syndrome

258. A bulimic 19-year-old female has a laceration in the lower esophagus. Which is the most probable condition?

a. hiatal hernia
b. diverticula
c. achalasia
d. Mallory-Weiss syndrome
e. peptic ulcer
f. cirrhosis
g. Gardener's syndrome
h. irritable bowel syndrome

259. Based on the previous question, which two of the following will further confirm your diagnosis?

a. arteriography
b. NCV
c. EMG
d. EEG
e. aspiration
f. myelography
g. bone scan
h. esophagoscopy
i. lateral lumbar view
j. RA latex

260. A 45-year-old male mortgage broker complains of dysphagia and chest discomfort when lying down after eating. X-rays shows a sac-like protrusion cephalad to the diaphragm. Which is the best choice?

a. esophageal cancer
b. heartburn
c. hiatal hernia
d. gastric ulcer
e. ascites
f. achalasia
g. acid reflux
h. esophageal webs

261. A 45-year-old male has trouble swallowing food. In addition, he has a pharyngeal pouches. Which is the most probable diagnosis?

a. sliding hiatal hernia
b. paraesophageal hiatal hernia
c. ascites
d. gastric ulcer
e. Zenker's diverticula
f. esophageal cancer
g. duodenal ulcer
h. acid reflux

262. A 54-year-old female complains of lumbar pain that is not relieved by sleep or exercise. An abdominal thrill is present upon palpation. X-rays reveal a curvilinear calcification. In addition, the temperature of the skin in the legs is cooler compared to the rest of the body. Which is the diagnosis?

a. abdominal aneurysm
b. acute pancreatitis
c. stomach cancer
d. sliding hiatal hernia
e. Zenker's diverticula
f. cauda equina syndrome
g. gastric ulcer
h. ascites

263. Based on the previous question, what are two other signs of this condition?

a. esophageal pouches
b. incontinence
c. constipation
d. Babinski's reflex
e. Gray-Turner's sign
f. Cullen's sign
g. mottled skin inferior to waist
h. pulsating epigastric mass

264. A 35-year-old female has an irregular-shaped, pinkish, nodular lesion on the back of her neck. This lesion also has a white border around it. She complains that this lesion sometimes bleeds occasionally. Her history reveals frequent sun-tanning as a teenage who used to smoke one pack of cigarettes a day. She also has herpes simplex type II. Which is the most probable condition?

a. Kaposi's sarcoma
b. dermatofibroma
c. keloid
d. basal cell carcinoma
e. multiple myeloma
f. malignant melanoma
g. seborric keratosis
h. actinic keratosis

265. Based on the previous question, which diagnostic tool would you use first to confirm this condition?

a. MRI
b. CAT scan
c. biopsy
d. x-ray
e. spirometer
f. NCV
g. tonometer
h. E.L.I.S.A.

266. A 34-year-old male has a fever of 105° F, chills, coughing and rust-colored sputum. Which is the diagnosis? Also, name the most common bacterial cause of this condition.

a. emphysema
b. pneumonia
c. bronchitis
d. sarcoidosis
e. TB
f. Streptococcus pneumoniae
g. Staphylococcus aureus
h. unknown
i. smoking
j. Legionella pneumophila

267. According to the previous question, which are two other principal signs of this condition?

a. shaking
b. chest pain
c. uveitis
d. glaucoma
e. erythema nodosum
f. migraine
g. Pott's disease
h. asthma
i. barrel chest
j. cavier lesions

268. Identify sounds heard in auscultation in the condition described in question 266?

a. wheezing
b. egophony
c. opening snap
d. whispered pectoriloquy
e. brochophony
f. wide splitting
g. reversed splitting
h. fixed splitting
i. resonant
j. hyperresonant

269. Based on question 266, which percussive note will be heard over affected area of the lung?
a. hyperresonant
b. resonant
c. dull
d. egophony
e. bronchophony
f. rales
g. wheezes
h. crackles

270. Which two of the following are used to test for malingering patients?
a. Magnuson's test
b. Allen's test
c. Apley's test
d. Ballottement test
e. Shrivel test
f. Cozen's test
g. Brudzinski sign
h. Burn's Bench test
i. Bikele's sign
j. Schepelmann's sign

271. A 39-year-old male personal trainer presents with diarrhea, abdominal pain and numbness in the legs. There is a loss of intrinsic factor. Which of the following best represents this condition?
a. sickle cell anemia
b. iron deficiency anemia
c. pernicious anemia
d. hepatitis A
e. hepatitis B
f. hepatitis C
g. pneumonia
h. bronchitis

272. Based on the previous question, which is the cause of this condition?
a. hereditary
b. IV drug use
c. A.I.D.S.
d. smoking
e. deficiency of B_{12}
f. deficiency of iron
g. sexual contact with infected partner
h. dirty food

273. A 40-year-old female has lost the taste of bitter. Which cranial nerve is most likely damaged?

a. 1
b. 2
c. 3
d. 4
e. 5
f. 7
g. 8
h. 9

274. A 3-year-old male presents with inflammation of the middle ear. There is an intense, constant earache coupled with bulging of the tympanic membrane in an outward direction. In addition, this patient also has the common cold. Which is the diagnosis?

a. acute otitis externa
b. acute otitis media
c. meningitis
d. vertigo
e. Meniere's disease
f. flu
g. otosclerosis
h. cholesteatoma

275. Based on question 274, what are two other physical characteristics of this condition?

a. diarrhea
b. scurvy
c. rickets
d. acne
e. psoriasis
f. Fordyce spots
g. fever of 105° F
h. anemia
i. hepatitis
j. torus palantinus

276. Based on question 274, which is the most common cause of this condition?

a. cranial nerve 8 injury
b. meningitis
c. osteomyelitis
d. bacteria entering by way of eustachian tube
e. cholesteatoma
f. hereditary
g. low blood iron levels
h. acute mastoiditis

277. A 23-year-old college swimmer complains of constant itching in the ear canal. Upon inspection, the inner ear is red and inflamed with foul-smelling pus. Which is the most likely condition?

a. exostosis
b. meningitis
c. cholesteatoma
d. osteomyelitis
e. otitis externa
f. otitis media
g. otosclerosis
h. Meniere's disease

278. Which are two causes of the previous condition described in question 277?

a. fungal infection
b. bacterial infection
c. perforated eardrum
d. Pancoast tumor
e. tonsillitis
f. peptic ulcer
g. ulcerative colitis
h. xerostomia
i. presbycusis
j. rickets

279. A 29-year-old male presents with a sense of fullness in the right ear. Upon inspection, the tympanic membrane is retracted. A yellowish-brown color is observed on the tympanic membrane. Which is the diagnosis?

a. otitis externa
b. bacterial otitis media
c. secretory otitis media
d. otosclerosis
e. presbycusis
f. cholesteatoma
g. exostosis
h. Pancoast tumor

280. A 49-year-old male complains of occasional muscle weakness and fatigue. Upon testing various muscle innervated by cranial nerves, muscle weakness is apparent. Which is the most likely condition?

a. multiple sclerosis
b. Parkinson's disease
c. myasthenia gravis
d. syringomyelia
e. subacute bacterial endocarditis
f. pleurisy
g. diabetes mellitus
h. hyperlipidemia

281. A 34-year-old female police officer complains of weight loss. Physical characteristics display exophthalmos, nervousness, and high energy. Which is the most probable condition? Also, name another factor involved with this condition.

a. Cushing syndrome
b. acromegaly
c. myxedema
d. Graves disease
e. SLE
f. moon face
g. decreased metabolism
h. increased metabolism
i. protruding jaw
j. loss of lateral eyebrows

282. Based on the previous question, which type of condition does this describe?

a. increased growth hormone
b. hypothyroidism
c. hyperthyroidism
d. cretinism
e. rheumatic fever
f. malignant melanoma
g. squamous cell carcinoma
h. lymphosarcoma

283. A 56-year-old female presents with yellow, dry, coarse skin, puffy eyes and loss of eyebrow hair. In addition, the person has gained a significant amount of weight in the last 6 months with decreased function of the thyroid gland. Which condition does this best describe?

a. Graves disease
b. acromegaly
c. psoriasis
d. SLE
e. myxedema
f. subacute bacterial endocarditis
g. cretinism
h. Addison's disease

284. A 42-year-old male has a suspected sacroiliac joint lesion. Which three of the following would you perform on this patient?

a. Allis test
b. Thomas test
c. Pelvic Rock test
d. Gaenslen's test
e. Bunnel Littler test
f. Ober test
g. Laguerre's test
h. Trendelenberg test
i. posterior apprehension test
j. Drawer's sign

285. Which are two ranges of motion for the ankle?

a. flexion
b. extension
c. abduction
d. elevation
e. retraction
f. adduction
g. external rotation
h. lateral flexion
i. pronation
j. supination

286. A college baseball pitcher has suspected medial epicondylitis. Flexion of the elbow is 95° and elbow extension is 4°. Name two other ranges of motion for the elbow.

a. adduction
b. abduction
c. internal rotation
d. external rotation
e. inversion
f. eversion
g. lateral flexion
h. supination
i. pronation
j. retraction

287. Based on question 286, which test would be most appropriate to perform on this patient?

a. Thomas test
b. Golfer's elbow test
c. Tinel's elbow sign
d. Hibb's test
e. Apley's scratch test
f. Shoulder Depression test
g. Dawbarn's sign
h. Bowstring sign

288. Vacuum phenomenon of Knuttson in the lumbar spine is found in a 54-year-old female. Decreased joint space between L3 and L4 is apparent on the lateral lumbar view. Which of the following is the most probable condition?

a. SLE
b. ankylosing spondylitis
c. Behcet's syndrome
d. psoriatic arthritis
e. osteoarthritis
f. Still's disease
g. Felty's syndrome
h. scleroderma

289. A 40-year-old self-employed, business woman presents with heartburn and constipation. There is marked swelling at the tip of the fingers. There appears to be multiple spider veins on the fingers and face. In addition, she has Raynaud's phenomenon. Which of the following is the diagnosis?

a. scleroderma
b. osteoarthritis
c. Felty's syndrome
d. psoriatic arthritis
e. Sever's disease
f. SLE
g. psoriatic arthritis
h. squamous cell carcinoma

290. Based on the previous question, what are two signs of this condition?

a. esophageal varices
b. protrusio acetabulae
c. os fabella
d. Herberden's nodes
e. Bouchard's nodes
f. Lover's heel
g. dry, silver scales
h. tophus
i. moles with irregular borders
j. dysphagia

291. Which of the following findings will usually be evident based on the description in question 289?

a. reversed A/G ratio
b. + ANA
c. elevated amylase
d. increased bilirubin
e. sero-positive
f. sero-negative
g. butterfly rash
h. decreased bilirubin
i. positive Reed Sternberg cells
j. decreased uric acid

292. Based on the diagnosis in question 289, which is not sign or symptom of this condition?

a. sclerodactyly
b. esophageal varices
c. calcinosis
d. dyspnea
e. Raynaud's phenomenon
f. telengetasia
g. floating stools
h. dagger sign

293. A 54-year-old grammar school teacher complains of intense pain and stiffness in the hips, neck and shoulders. She also notes that it feels worse in the morning or when she doesn't exercise for a whole day. ESR is elevated. Which is the most likely condition?

a. AS
b. Wegener's granulomatosis
c. Reiter's syndrome
d. polymyalgia rheumatica
e. gout
f. osteomyelitis
g. Behcet's syndrome
h. psoriatic arthritis

294. A 23-year-old male college student presents with painful mouth sores. In addition, there are skin blisters and pimples on the trunk, back and facial areas. This patient also complained that his skin has become very inflamed around the area where he recently got a flu shot. Which is the diagnosis?

a. scleroderma
b. multiple myeloma
c. Reiter's syndrome
d. Behcet's syndrome
e. osteomyelitis
f. syphilis
g. meningitis
h. Wegener's granulomatosis

295. Based on the previous question, what are two other physical signs or symptoms of this condition?

a. vasculitis
b. positive Kernig's test
c. rain drop skull
d. floating stools
e. pencil-in-cup deformity
f. gummas
g. iridocyclitis
h. tabes dorsalis
i. Lover's heel
j. esophageal varices

296. A 56-year-old male has decreased sensation in his wrist and ankle joints. In fact, he has trouble feeling pain in these joints. Which of the following best describes this condition?

a. meningitis
b. osteoarthritis
c. RA
d. Reiter's syndrome
e. torticollis
f. scleroderma
g. Charcot's joints
h. multiple myeloma

297. Based on the diagnosis in the previous question, which three conditions does this occur in?

a. diabetes mellitus
b. syphilis
c. leprosy
d. golfer's elbow
e. otosclerosis
f. lateral epicondylitis
g. osteoma
h. osteoid osteoma
i. Meniere's disease
j. carpal tunnel syndrome

298. While testing a reflex on a 34-year-old female patient, the toes curl. This occurs when stroking the sole of the foot with the tip of a reflex hammer handle. Which reflex does this describe?

a. Hoffman's
b. Babinski's
c. Oppenheim's
d. Mayer's
e. Schaefer's
f. Plantar
g. Cremaster
h. Klippel-Weil

299. A 29-year-old male has been diagnosed with multiple renal calculi. Identify three signs or symptoms usually observed in this condition.

a. biliary atresia
b. intussusception
c. hematuria
d. flank pain
e. substernal pain
f. migraine
g. gastroesophageal reflux
h. spider veins
i. fever
j. neuroblastoma

300. A 52-year-old male accountant presents with constipation, left lower quadrant pain and nausea. Upon abdominal auscultation, decreased bowel sounds are apparent. Which is the most likely diagnosis?

a. sliding hiatal hernia
b. diverticulitis
c. kidney stones
d. biliary atresia
e. Wilms' tumor
f. neuroblastoma
g. Hirschsprung disease
h. acid reflux

301. A 42-year-old female is diagnosed with breast cancer. Identify two physical signs of this condition.

a. fever
b. migraine
c. peau d'orange
d. hiatal hernia
e. painless mass
f. gynecomastia
g. kidney stones
h. galactorrhea
i. mastitis
j. mammary duct ectasia

302. A 22-year-old male involved in a car accident presents with muffled heart sounds, paradoxic pulse and low blood pressure. Which is the most likely condition? Also, name another cause of this condition.

a. MI
b. cardiac tamponade
c. bacterial endocarditis
d. mitral stenosis
e. coarctation of the aorta
f. pericarditis
g. Raynaud phenomenon
h. angina
i. aortic sclerosis
j. tetralogy of Fallot

303. After performing the Klippel-Weil reflex, which finding can be labeled pathological on the patient?

a. Babinski response
b. extension of fingers
c. extension of thumb
d. knee extension
e. flexion of big toe
f. curling of toes
g. adduction and flexion of thumb
h. dorsiflexion of foot

304. After evaluating the L4 dermatome, which muscle is part of the L4 neurological package?

a. peroneus longus
b. rectus abdominus
c. gluteus maximus
d. peroneus brevis
e. iliopsoas
f. tibialis anterior
g. extensor digitorum longus
h. extensor digitorum brevis

305. A 23-year-old male presents with low back pain that is worse during the late night hours. Mornings usually begin with back stiffness. ESR is

elevated and kyphotic deformity seems to be developing. Based on this information, which is the most probable condition?

a. RA
b. osteoarthritis
c. ankylosing spondylitis
d. Still's disease
e. scleroderma
f. meningitis
g. Paget's disease
h. lymphosarcoma

306. Which of the following two would you perform to further confirm this diagnosis?

a. Forestier's Bowstring sign
b. McMurray sign
c. Thomas test
d. Kemp's test
e. Brudzinski test
f. Bakody test
g. Percussion test
h. Chest Expansion test
i. LHermitte's sign
j. Schepelmann's sign

307. Based on question 305, which are two physical signs of this condition?

a. migraine
b. brain tumor
c. increase in hat size
d. Haygarth's nodes
e. fatigue
f. dermatitis
g. weight loss
h. sclerodactyly
i. Raynaud's phenomenon
j. urethritis

308. Based on question 305, what are two signs of this condition on x-ray? Also, identify a lab finding associated with this condition.

a. Bamboo spine
b. shiny corner sign
c. cocktail sausage digit
d. podagra
e. Bouchard's node
f. RA latex (+)
g. (+) LE prep
h. (+) FANA
i. uric acid
j. (+) HLA-27

309. A 62-year-old female presents with deep pain in the muscles and bones that get worse with standing and decreases by lying supine. Upon x-ray evaluation, there are numerous punched out lesions throughout the ribs and ilium. The lesions are less than 2 cm in diameter. Which is the most probable condition?

a. osteoma
b. RA
c. scleroderma
d. multiple myeloma
e. osteoporosis
f. osteomalacia
g. chordoma
h. chondrosarcoma

310. Based on question 309, which are three findings observed with this condition?

a. (+) Protein electrophoresis
b. increased bilirubin
c. (+) Le prep
d. decreased bone density
e. hypercalcemia
f. positive Brudzinski test
g. elevated GGT
h. elevated HCG
i. vitiligo
j. blebs

311. Based on question 309, which are three other physical signs or symptoms of this condition?

a. weight loss
b. mastitis
c. ulcerative colitis
d. apthous ulcer
e. Fordyce spots
f. night sweats
g. pityriasis rosea
h. Kerr's sign
i. weakness
j. loss of lateral one-third of eyebrow

312. A nine-year-old male presents with restlessness and irritability. Rachitic rosary is also evident. In addition, bowing deformity of the femur is also observed on x-ray. Which is the diagnosis?

a. osteoporosis
b. Still's disease
c. rickets
d. osteosarcoma
e. multiple myeloma
f. Ewing's tumor
g. TB
h. osteomyelitis

313. Based on the previous question, which is a lab finding associated with this condition. Also, identify another radiographic sign in this condition.

a. elevated alkaline phosphatase
b. (+) RA latex
c. uric acid
d. (+) Le prep
e. decreased lipase
f. rain drop skull
g. Schmorl's nodes
h. veterbral plana
i. trumpeting
j. dagger sign

314. Which of the following conditions will amylase levels be elevated?

a. multiple myeloma
b. Sudeck's atrophy
c. acute pancreatitis
d. scleroderma
e. Reiter's syndrome
f. RA
g. rickets
h. Still's disease

315. A 39-year-old female has been diagnosed with hyperthyroidism. Identify 3 signs or symptoms that occur in this disease.

a. vitamin D deficiency
b. elevated SGOT
c. cirrhosis
d. emphysema
e. goiter
f. increased bowel movements
g. migraine
h. nervousness
i. decreased bowel movements
j. bradycardia

316. An overweight, 41-year-old woman complains of colicky pain in the upper right quadrant. She says this happens most often after eating a heavy meal, which consists of a lot of dairy. In addition, her urine is black-brown in color. Which is the most likely condition?

a. gastric carcinoma
b. renal calculi
c. sliding hiatal hernia
d. appendicitis
e. acid reflux
f. diverticulitis
g. cholestasis
h. ulcerative colitis

317. Based on the previous question, which is one cause of this condition?

a. congenital malformation
b. stress
c. Scarlet fever
d. gallstones
e. ulcerative colitis
f. stasis dermatitis
g. neurapraxia
h. melanoma

318. A 39-year-old carpenter has been informed to perform Codman's exercises by his family doctor. Which of the following is a possible diagnosis?

a. ankle strain
b. ankle sprain
c. cervical whiplash
d. frozen shoulder syndrome
e. fractured tibia
f. Sever's disease
g. heel spur
h. 20 degree scoliosis

319. A 35-year-old female has been diagnosed with multiple sclerosis. Which of the following can further confirm this diagnosis?

a. Phalens test
b. Lhermitte's sign
c. O'Donoghe maneuver
d. Lewin's test
e. Sicard's test
f. Straight Leg Raiser
g. Well Leg Raiser
h. Ely's sign

320. A 49-year-old male has been diagnosed with ptosis and decreased accommodation reflex. Which two of the following are possibly damaged?

a. olfactory nerve
b. optic nerve
c. trigeminal nerve
d. facial nerve
e. occulomotor nerve
f. vestibulo-acoustic nerve
g. spinal accessory nerve
h. vagus nerve
i. glossopharyngeal nerve
j. hypoglossal nerve

321. Buffalo hump, moon face and excessive hair growth is apparent in a 45-year-old female. Which of the following is the most likely condition?

a. myxedema
b. acromegaly
c. Huntington's chorea
d. Cushing's syndrome
e. multiple sclerosis
f. tuberculosis
g. Sydenham's chorea
h. scleroderma

322. A 47-year-old male presents with tiny, irregular pupils that don't react to light. However, the accommodation reflex is present. Which of the best describes this condition of the eye?

a. anisocoria
b. entropion
c. extropion
d. xanthelasma
e. Keyser-Fleischer ring
f. pinguecula
g. chalazion
h. Argyll-Robertson pupil

323. Based on the condition described in question 322, which diagnosis should be suspected?

a. Horner's syndrome
b. Huntington's chorea
c. syphilis
d. herpes simplex type I
e. myxedema
f. Addison's disease
g. acromegaly
h. venereal warts

324. A 32-year-old woman is diagnosed with xanthelasma right above the eyelid. Which is the most probable diagnosis?

a. multiple sclerosis
b. Parkinson's disease
c. abnormality of lipid metabolism
d. syphilis
e. herpes simplex type II
f. shingles
g. Addison's disease
h. Grave's disease

325. A 58-year-old retired police officer has been diagnosed with diabetic retinopathy. Identify three findings observed in this condition.

a. Keyser-Fleischer ring
b. soft exudates
c. cataract
d. chalazion
e. dot hemorrhages
f. acute hordeolum
g. pinguecula
h. anisocoria
i. hard exudates
j. Argyll-Robertson Pupil

326. A 45-year-old male has been diagnosed with Erb's paralysis. Which nerve root is affected? Also, which is a characteristic of this condition.

a. C1-C2
b. C2-C3
c. C4
d. C5-C6
e. C8-T1
f. waiter's tip
g. claw hand
h. Charcot's triad
i. dysarthria
j. adhesive capsulitis

327. A 52-year-old male presents with a monotone voice, resting tremor, and festinating gait. Which is the cause of this condition?

a. parotid gland enlargement
b. hypothyroidism
c. hyperthyroidism
d. adrenal cortex dysfunction
e. dopamine insufficiency
f. hyperthyroidism
g. demyelination of CNS
h. wearing away of caudate nucleus

328. Based on the previous question, name two other physical signs of this condition.

a. cogwheel rigidity
b. intention tremor
c. Lhermitte's sign
d. oily skin
e. torticollis
f. athetoid movements
g. goiter
h. inflammation near ears
i. loss of lateral 1/3 of eyebrow
j. pinguecula

329. According to questions 327 and 328, which is the diagnosis?

a. multiple sclerosis
b. myxedema
c. Cushing's syndrome
d. parotid gland enlargement
e. Huntington's chorea
f. Graves disease
g. Parkinson's disease
h. meningitis

330. A 23-year-old college student complains of posterior neck soreness. Which of the following can further distinguish this to be either a strain or sprain?

a. Lindner's test
b. Brudzinski test
c. O'Donoghue maneuver
d. Lhermitte's sign
e. Bakody sign
f. Adson's test
g. Compression test
h. Maigne's test

331. A 45-year-old male patient shows a Babinski response after squeezing the Achilles tendon. Which pathological reflex does this describe?

a. Chaddock's
b. Oppenheim's
c. Gordon's
d. Hoffman's
e. Tromner's
f. Ankle clonus
g. McCarthy's
h. Schaffer's

332. The gluteus maximus muscle is given a muscle grade of 1 in a 45-year-old male patient. Which neurological level is the gluteus maximus tested under?

a. L1
b. L2
c. L3
d. L4
e. L5
f. S1
g. C5
h. T10

333. Based on the previous question, identify the action to test the gluteus maximus. Also, identify the reflex that should be tested for this neurological level?

a. foot dorsiflexion
b. hip extension
c. foot inversion
d. leg abduction
e. hip flexion
f. Achilles reflex
g. patella reflex
h. triceps reflex
i. biceps reflex
j. Babinski reflex

334. A 40-year-old male presents with athetosis and occasional purposeless,

rapid movements. Memory loss is also apparent. Which is the most likely condition?

a. Huntington's disease
b. multiple sclerosis
c. Parkinson's disease
d. myasthenia gravis
e. thoracic outlet syndrome
f. cervical spondylosis
g. sarcoidosis
h. lymphocytic leukemia

335. According to the previous question, what are two signs of this condition?

a. dementia
b. loss of hearing
c. intention tremor
d. resting tremor
e. migraine
f. anemia
g. blebs
h. osteophyte
i. dysarthria
j. cervical rib

336. Based on the question 334, what is the cause of this condition?

a. disc herniation
b. dopamine deficiency
c. demyelination of CNS
d. copper excess
e. meningitis
f. wearing away of caudate nucleus
g. emphysema
h. cervical rib
i. thoracic outlet syndrome
j. aluminum poisoning

337. A 30-year-old male has a suspected dislocated shoulder. Which of the following will most likely be positive? (Identify two choices.)

a. Halstead maneuver
b. Cozen's test
c. Kaplan's sign
d. Apprehension test
e. Roos test
f. Shrivel test
g. Bracelet test
h. Dugas test
i. McMurray sign
j. Thomas test

338. A 35-year-old male presents with a suspected chondromalacia patella. Which of the following will be positive?

a. anvil test
b. Ober's test
c. Clarke's sign
d. Beevor's sign
e. Minor's sign
f. Thomas test
g. Laguerre's test
h. Belt test

339. A 72-year-old woman has been diagnosed with osteoporosis in the

cervical, thoracic and lumbar spine. Hyperkyphosis is apparent. Which are two findings of this condition?

a. McConkey's sign
b. cumulus cloud lesion
c. chordoma
d. aortic aneurysm
e. peptic ulcer
f. biconcave vertebrae
g. Pott's disease
h. tension headache
i. torticollis
j. Grey-Turner's sign

340. Picture frame vertebra, brim sign and shepherd's crook are apparent in a 69-year-old male. Deep, aching bone pain is worse at night. Which is the most probable condition?

a. multiple myeloma
b. hematoma
c. rheumatoid arthritis
d. osteoporosis
e. Paget's disease
f. osteopoikilosis
g. Klippel-Feil syndrome
h. osteoarthritis

341. Based on question 340, which of the following lab findings will be elevated?

a. lipase
b. amylase
c. alpha-fetoprotein
d. alkaline phosphatase
e. creatinine clearance
f. glucose
g. SGOT
h. cholesterol

342. Based on question 340, which are two other radiographic findings observed in this condition?

a. low hairline
b. Sprengel's deformity
c. cotton wool skull
d. protrusio acetabuli
e. short, webbed neck
f. punched out lesions
g. rain drop skull
h. osteoma
i. Salter-Harris Type IV fracture
j. vertebra plana

343. A 34-year-old male presents with weakness in the arms and feet. Muscle spasms are also common. In addition, dysarthria and dysphagia are present. Which of the following is the most probable condition?

a. amyotrophic lateral sclerosis
b. multiple sclerosis
c. Paget's disease
d. osteoma
e. osteoporosis
f. Parkinson's disease
g. unicameral bone cyst
h. osteomyelitis

344. What are two other possible physical signs of the condition described in question 343?

a. Adie's pupil
b. cellulits
c. Baker's cyst
d. hydrocele
e. hyperreflexia of the legs
f. paralysis
g. mastitis
h. Murphy's sign
i. pityriasis rosea
j. Charcot's triad

345. A 52-year-old male alcoholic presents with nystagmus, double vision and amnesia. In addition, this patient complains of being drowsy. Which of the following is the most probable diagnosis?

a. Huntington disease
b. Alzheimer disease
c. Wernicke-Korsakoff syndrome
d. meningitis
e. diabetes mellitus type II
f. diabetic retinopathy
g. tabes dorsalis
h. hordeolum

346. Based on question 345, which of the following is a cause of this condition?

a. aluminum poisoning
b. thiamine deficiency
c. Neisseria meningitidis
d. Streptococcus pneumoniae
e. Hemophilus influenzae
f. Kaposi's sarcoma
g. thalassemia
h. hypothyroidism

347. A 32-year-old male presents with a runny nose and watery eyes. He complains that there is moderate to severe pain behind the eye. This pain occurs in episodes of 4 times per day. This has been happening for the past 6 weeks. Which of the following is the most likely diagnosis?

a. brain tumor
b. hypertensive headache
c. tension headache
d. brain abscess
e. cluster headache
f. classic migraine
g. common migraine
h. osteoma

348. According to the previous question, which are two factors that can bring on this condition?

a. high altitudes
b. Charcot's triad
c. disc herniation
d. multiple sclerosis
e. vitamin A deficiency
f. brain tumor
g. diverticulum
h. hiatal hernia
i. spina bifida
j. alcohol consumption

349. A 31-year-old male presents with pitted nails. In addition, cocktail sausage digit is observed on x-rays. Lab reveals negative rheumatoid factor and elevated ESR. Which is the diagnosis?

a. ankylosing spondylitis
b. enteropathic arthritis
c. Reiter's syndrome
d. scleroderma
e. osteosarcoma
f. osteoarthritis
g. rheumatoid arthritis
h. psoriatic arthritis

350. Based on question 349, what are two radiographic signs of the condition?

a. punched out lesions
b. trumpeting
c. bowing deformity
d. osteoma
e. osteonecrosis
f. Codman's triangle
g. gull wings
h. shiny corner sign
i. Romanus lesion
j. mouse ears

351. Based on question 349, which of the following radiographic signs is practically diagnostic of this condition?

a. ray pattern
b. ivory vertebra
c. loss of joint space
d. moth-eaten appearance
e. Haygarth's nodes
f. Heberden node
g. ulnar deviation
h. shiny corner sign

352. Based on question 349, which are two radiographic appearances commonly seen in the spine or sacrum with this condition?

a. soap bubble lesion
b. osteophytes
c. cumulus cloud appearance
d. centralized flake of calcification
e. non-marginal syndesmophyte
f. fallen fragment sign
g. picture frame vertebra
h. vertebra plana
i. bilateral sacroiliitis
j. gull wings

353. Based on question 349, which is the closest differential diagnosis to

this condition on spinal radiographs?

a. SLE
b. osteosarcoma
c. DJD
d. Reiter's syndrome
e. psoriatic arthritis
f. gout
g. C.P.P.D.
h. osteoma

354. Non-marginal syndesmophytes are apparent in the lumbar spine of a 38-year-old male with a history of dysentery. HLA-B27 is positive. A foot spur is also apparent on x-rays. Which is the most probable condition?

a. AS
b. SLE
c. gout
d. pseudogout
e. Reiter's syndrome
f. RA
g. scleroderma
h. multiple myeloma

355. Based on the condition in the previous question, what are two clinical signs of this condition?

a. migraine
b. polyarthritis
c. conjuctivitis
d. wrist drop
e. butterfly rash
f. torticollis
g. enchondroma
h. femoral hernia
i. Froment's sign
j. Kerr's sign

356. After evaluating a 45-year-old male patient's trigeminal nerve, muscle weakness is apparent. Which is one muscle that has been tested? Also, name the reflex associated with this cranial nerve?

a. SCM
b. trapezius
c. masseter muscle
d. supraspinatus
e. soleus
f. gag reflex
g. jaw jerk reflex
h. uvular reflex
i. plantar reflex
j. Hoffman reflex

357. A 32-year-old woman presents with a rash on the elbows and hands. There is pain and stiffness in the wrist and hands. Radiographs show boutonniere deformity of the 1st metacarpal bilaterally. Which are three lab findings in this condition?

a. elevated acid phosphatase
b. low triglycerides
c. elevated ESR
d. (+) LE cell
e. decreased lipase
f. decreased SGOT
g. elevated direct bilirubin
h. elevated PSA
i. elevated Anti-nuclear Antibody
j. elevated A.S.O. titers

358. Based on question 357, what are two other signs of this condition?
a. reversible bone deformity
b. dagger sign
c. bamboo spine
d. Lover's heel
e. atypical syndesmophytes
f. mouse ears
g. gullwings
h. osteoporosis
i. ray pattern
j. cocktail sausage digit

359. Based on question 357 and 358, which of the following is the diagnosis?
a. AS
b. Reiter's syndrome
c. psoriatic arthritis
d. osteoarthritis
e. Systemic Lupus Erythematosus
f. gout
g. osteitis pubis
h. scleroderma

360. A 27-year-old female presents with abdominal pain and weight loss. X-ray evaluation reveals marginal syndesmophytes in the lumbar spine. Also, bilateral sacroiliitis is apparent. This person has a history of Crohn's disease. Which is the most probable condition?
a. RA
b. DJD
c. psoriatic arthritis
d. enteropathic arthritis
e. AS
f. Systemic Lupus Erythematosus
g. Reiter's syndrome
h. scleroderma

361. Based on question 360, which is the closest differential diagnosis of this condition based on the following list?
a. DJD
b. RA
c. AS
d. SLE
e. scleroderma
f. Reiter's syndrome
g. enteropathic arthritis
h. osteosarcoma

362. Based on the answer in 360, which is a condition that causes this diagnosis?

a. scleroderma
b. ulcerative colitis
c. osteoarthritis
d. osteosarcoma
e. osteoid osteoma
f. Reiter's syndrome
g. diabetes mellitus
h. hematoma

363. A 47-year-old male complains of an inability to jog or run. X-ray evaluation shows knee joint debris, disorganization and bone destruction. Which of the following is the most likely diagnosis?

a. RA
b. psoriatic arthritis
c. neuropathic arthritis
d. scleroderma
e. osteoma
f. hematoma
g. SLE
h. sarcoidosis

364. A 27-year-old male presents with stiffness and pain in the lumbar and sacral spine that is worse in the early morning hours. In addition, this patient is 10 pounds underweight. There is also limited chest expansion when taking deep breaths. Which is the most likely diagnosis?

a. Reiter's syndrome
b. psoriatic arthritis
c. RA
d. scleroderma
e. lymphosarcoma
f. AS
g. DJD
h. sarcoidosis

365. Based on the previous question, identify three radiographic signs observed in this condition?

a. carrot stick fracture
b. marginal syndesmophyte
c. star sign
d. Lover's heel
e. mouse ears
f. pencil-in-cup deformity
g. cocktail sausage digit
h. calcific lymphadenopathy
i. hilar enlargement
j. blebs

366. A 20-year-old field hockey player presents with degeneration of the tibial tuberosity. Which is the most likely condition?

a. RA
b. AS
c. scleroderma
d. juvenile RA
e. Osgood Schlatter's disease
f. Scheuermann's disease
g. Legg Calve Perthes disease
h. sarcoidosis

367. Stenosis of the basilar artery is present in a 53-year-old female.

Identify three of the following that will be positive.

a. Soto Hall test
b. Lindner's test
c. Brudzinski test
d. Dekleyn's test
e. Percussion test
f. Turyn's test
g. Hautant's test
h. Barre-Lieou sign
i. Ely's sign
j. Bonnet's sign

368. A 32-year-old grammar school teacher presents with fever, fatigue and a headache. She also complains of knee and shoulder pain that she has never had before. Erythema chronicum migrans is also apparent on the anterior forearm. Which is the most likely condition?

a. psittacosis
b. syphilis
c. measles
d. scleroderma
e. psoriasis
f. Lyme disease
g. Scarlet fever
h. toxic shock syndrome

369. Based on the diagnosis in question 368, what are two other clinical findings of this condition?

a. arrhythmia
b. Adie's pupil
c. duodenal ulcer
d. gastric ulcer
e. hepatitis
f. hiatal hernia
g. stiff neck
h. pancreatitis
i. diabetes
j. oral thrush

370. A 48-year-old female presents with puffy eyelids, hypercholesterolema and proteinuria. In addition, there are decreased levels of blood albumin. Which is the most likely diagnosis?

a. pancreatitis
b. hepatitis
c. mononucleosis
d. Graves disease
e. nephrotic syndrome
f. thyrotoxicosis
g. syphilis
h. amyotrophic lateral sclerosis

371. A 37-year-old female complains of constantly having dry eyes and dry mouth. She also states that she has had three cavities filled in the last four months. Lab reveals an elevated ESR. Which is the diagnosis? Also, name another symptom or sign of this condition.

a. scleroderma
b. osteoarthritis
c. AS
d. tuberculosis
e. Sjogren's syndrome
f. dagger sign
g. Raynaud's phenomenon
h. floating stools
i. Fordyce spots
j. joint inflammation

372. A 21-year-old male presents with stick neck, chills and occasional vomiting. Kernig's test is positive. Which two other tests would you perform to further confirm your diagnosis?

a. O'Donoghue maneuver
b. Lindner's test
c. Bakody sign
d. Percussion test
e. Distraction test
f. Adam's sign
g. Adson's test
h. Brudzinski test
i. Patrick test
j. Anvil test

373. Based on question 372, which is the diagnosis?

a. hip joint lesion
b. sciatica
c. scoliosis
d. meningitis
e. multiple sclerosis
f. thoracic outlet syndrome
g. disc herniation
h. spinal fracture

374. A 68-year-old male presents with dysphagia. Lateral cervical radiographs reveals calcification of the anterior longitudinal ligament from C3-C7. History, reveals diabetes mellitus type II. Which is the diagnosis?

a. AS
b. RA
c. DJD
d. osteoma
e. tuberculosis
f. pseudogout
g. psoriatic arthritis
h. D.I.S.H.

375. Based on question 374, which is another symptom or sign usually present in this condition?

a. morning pain and stiffness
b. migraine
c. meningitis
d. saber shin
e. tremors
f. Volkmann's contracture
g. Kerr's sign
h. Morton's neuroma

376. Identify two terms related to the condition described in question 374.

a. overhanging edge sign
b. Romanus lesion
c. mouse ears
d. Haygarth's nodes
e. dagger sign
f. pericarditis
g. dripping candle wax
h. ivory vertebra
i. blade of grass
j. flame shaped ostephyte

377. Identify two differential diagnoses for the condition in question 374.

a. AS
b. DJD
c. gout
d. pseudogout
e. osteomyelitis
f. scleroderma
g. lymphosarcoma
h. tuberculosis
i. Paget's disease
j. non-ossifying fibroma

378. A 15-year-old female has a structural scoliosis. Which of the following may further confirm this diagnosis?

a. Beevor's sign
b. Phalens test
c. Adam's sign
d. Adson's test
e. Allen test
f. Halstead test
g. O'Donoghue maneuver
h. Soto Hall test

379. A 12-year-old female presents with peroneal muscular atrophy. Foot drop is apparent with atrophy of the calf muscles. Which is the diagnosis?

a. Charcot-Marie-Tooth disease
b. Scarlet fever
c. Sydenham's chorea
d. multiple sclerosis
e. thrombophlebitis
f. tarsal tunnel syndrome
g. Klippel-Feil syndrome
h. Scheuermann's disease

380. Based on the previous question, which of the following is a physical sign of this condition?

a. Kerr's sign
b. trolley track spine
c. stork leg deformity
d. tophus
e. dysphagia
f. Fordyce spots
g. Charcot's triad
h. aphasia

381. A 48-year-old male has been diagnosed with neurotrophic arthropathy. Name two findings observed in this condition.

a. Charcot's joints
b. hiatal hernia
c. Lover's heel
d. jigsaw vertebra
e. cocktail sausage digit
f. tophus
g. Bouchard's nodes
h. spina bifida occulta
i. knife clasp deformity
j. posterior ponticus

382. Which two conditions can cause neurotrophic arthropathy?
a. Parkinson's disease
b. Scarlet fever
c. peptic ulcer
d. pyloric steonsis
e. syphilis
f. chronic bronchitis
g. osteoma
h. osteoid osteoma
i. roseola infantum
j. diabetes mellitus type II

383. Baker's cysts, Hitchhiker's thumb, swan-neck deformity and Lanois deformity are observed in a 30-year-old female college professor. Which of the following is the most probable diagnosis?
a. DJD
b. RA
c. AS
d. Os Odontoidium
e. spondylolisthesis
f. tuberculosis
g. fibrous dysplasia
h. osteosarcoma

384. A 34-year-old multiparous female presents with pain and stiffness in the areas of L3-S2. There are triangular in shape regions of sclerosis on the bilateral inferior portions of the ilii. Which is the most accurate diagnosis?
a. AS
b. RA
c. osteitis pubis
d. osteitis ilii
e. DJD
f. scleroderma
g. hypertrophic osteoarthropathy
h. Forestier's disease

385. A 70-year-old male with a history of bronchogenic carcinoma presents with clubbing of the fingers, pain in the leg and inflammation of the elbow and knee joints. Which is the most likely condition?
a. AS
b. tuberculosis
c. scleroderma
d. hypertrophic osteoarthropathy
e. fibrous dysplasia
f. osteoma
g. osteosarcoma
h. gout

386. Based on the previous question, which is another radiographic sign of

this condition?

a. periostitis
b. tophus
c. dagger sign
d. Pott's disease
e. cumulus cloud appearance of bone
f. Pie sign
g. bone island
h. rain drop skull

387. A 36-year-old woman presents with an antalgic gait. X-rays reveal subcondral sclerosis and osteoporosis around the symphysis pubis. Which is the diagnosis?

a. osteitis ilii
b. Forestier's disease
c. osteitis pubis
d. gout
e. fibrous dysplasia
f. scleroderma
g. osteoarthritis
h. AS

388. Based on the previous question, which are two causes for this condition in a male or female?

a. gout
b. hiatal hernia
c. ulcerative colitis
d. basilar invagination
e. bronchiectasis
f. angina pectoris
g. hypothyroidism
h. platybasia
i. prostate surgery
j. pregnancy

389. Based on question 387, which is another radiographic sign of this condition?

a. widened joint space
b. Haygarth nodes
c. tophus
d. acro-osteolysis
e. Lover's heel
f. shepherd's crook
g. ivory vertebra
h. fallen fragment sign

390. A 50-year-old male has been diagnosed with gout. Which are three clinical aspects of this disease?

a. elevated lipase
b. tophi
c. leukocytosis
d. decreased acid phosphatase
e. increased ESR
f. Brodie's abscess
g. picture frame vertebra
h. chondroblastoma
i. shepherd's crook
j. soap bubble lesion

391. Identify two radiographic signs observed in gout.

a. osteoma
b. atypical syndesmphytes
c. typical syndesmophytes
d. shiny corner sign
e. spotty carpal sign
f. punched-out lesions
g. moth-eaten lesions
h. overhanging margin sign
i. snowflake calcification
j. soap bubble lesion

392. A 24-year-old Scandinavian female has been diagnosed with sarcoidosis. Which is the most common site of occurrence? Also, describe the type of lesion found in this condition.

a. cervical spine
b. shoulder
c. hands
d. feet
e. ribs
f. radiolucent bone lesions
g. soap bubble lesions
h. cumulus cloud appearance
i. popcorn calcification
j. centralized flake of calcification

393. Based on the previous question, which is the cause for the type of bone lesion observed in sarcoidosis?

a. osteosarcoma
b. multiple myeloma
c. chondrosarcoma
d. Ewing's sarcoma
e. granuloma
f. osteoma
g. osteoid osteoma
h. enchondroma

394. Based on question 392, what are two signs or symptoms of this condition?

a. migraine
b. erythema nodosum
c. Fordyce spots
d. weight loss
e. pigeon chest deformity
f. coxa vara
g. positive Turyn's test
h. protusio acetabuli
i. intention tremor
j. temporal arteritis

395. Biconcave vertebra, vertebral plana and Schmorl's nodes are evident on x-rays of a 70-year-old male. McConkey's sign is also observed. Which is the most likely diagnosis?

a. AS
b. RA
c. osteoarthritis
d. osteoporosis
e. gout
f. scleroderma
g. multiple myeloma
h. condrosarcoma

396. A 68-year-old male presents with a forward, slumped posture coupled with a resting tremor. Drooling is also present with a facial mask-like stare. Which is the diagnosis?

a. Huntington's chorea
b. multiple sclerosis
c. A.L.S.
d. gout
e. Tay-Sachs disease
f. Parkinson's disease
g. nephrotic syndrome
h. Cushing's syndrome

397. A 54-year-old female presents with a thick tongue, loss of lateral 1/3 of eyebrow and dry, rough hair. Which is the most probable condition?

a. Cushing's syndrome
b. myxedema
c. multiple sclerosis
d. RA
e. hyperthyroidism
f. Huntington's chorea
g. Paralysis Agitans
h. Lou Gehrig's disease

398. Based on the previous question, which are two other physical signs of this condition?

a. carotenemia of palms
b. intention tremor
c. increased metabolism
d. increased appetite
e. tachycardia
f. nervousness
g. dull facial expression
h. hirsutism
i. moon face
j. swan-neck deformity

399. A 34-year-old male presents with pigeon breast deformity, nystagmus and dysarthria. Clinical evidence reveals cape-like sensory deficient. Which is the most likely condition?

a. multiple myeloma
b. multiple sclerosis
c. Parkinson's disease
d. myasthenia gravis
e. syringomyelia
f. hyperthyroidism
g. hypothyroidism
h. Lou Gehrig's disease

400. A 33-year-old woman presents with ptosis of the eyelids. She also complains of excessive fatigue after moderate exercise. Other symptoms include difficulty speaking and dysphagia. Which is the diagnosis?

a. syringomyelia
b. RA
c. myasthenia gravis
d. AS
e. Cushing's syndrome
f. Paralysis Agitans
g. C.P.P.D.
h. hiatal hernia

Technique Practical

The technique portion of this exam is based slightly on word play. This word play is given in the form of Medicare listings, Gonstead (Palmer) listings or on rare occasions National listings. Usually, vertebral listings are given in Medicare and Palmer (Gonstead) listings. Depending on what chiropractic college you graduated from, you may not be familiar with Medicare listings. The clues to the listings are given before each adjusting room. You must know how to figure out the listing in order to perform the correct adjustment.

Medicare listing- refers to the vertebral body as a reference.

1. Medicare listing of left rotational malposition, left lateral flexion malposition is equivalent to a PRS-SP Gonstead (Palmer) listing.
2. Medicare listing of left rotational malposition, right lateral flexion malposition is equivalent to a PRI-L, PRI-T or PRI-L Gonstead (Palmer) listing.
3. Medicare listing of right rotational malposition, right lateral flexion malposition is equivalent to a PLS-SP Gonstead (Palmer) listing.
4. Medicare listing of right rotational malposition, left lateral flexion malposition is equivalent to a PLI-L, PLI-T or PLI-M Gonstead (Palmer) listing.
5. Medicare listing of left rotational malposition is equivalent to a PR Gonstead (Palmer) listing.
6. Medicare listing of right rotational malposition is equivalent to a PL Gonstead (Palmer) listing.

Gonstead (Palmer) listings- use the spinous process as the reference.

1. PRS-SP is usually described as posterior right, superior spinous.
2. PRI-L, PRI-T or PRI-M is usually described as posterior right, inferior spinous.
3. PLS-SP is usually described as posterior left, superior spinous.
4. PLI-L, PLI-T or PLI-M is usually described as posterior left, inferior spinous.
5. PR is usually described as posterior spinous right.
6. PL is usually described as posterior spinous left.

National listings- use the vertebral body as the reference.

1. National listing of left posterior inferior is equivalent to a PRS-SP Gonstead (Palmer) listing.
2. National listing of left posterior superior is equivalent to a PRI-L, PRI-T or PRI-M Gonstead (Palmer) listing.
3. National listing of right posterior inferior is equivalent to a PLS-SP Gonstead (Palmer) listing.
4. National listing of right posterior superior is equivalent to a PLI-L, PLI-T or PLI-M Gonstead (Palmer) listing.
5. National listing of left posterior is equivalent to a PR Gonstead (Palmer) listing.
6. National listing of right posterior is equivalent to a PL Gonstead (Palmer) listing.

More important facts to remember.

Side of lateral flexion restriction is the side of the open wedge. So if there is a right lateral flexion restriction at T6, the right side is the location of the open wedge.

-Right lateral flexion restriction means the right side is the side of the open wedge.

-Left lateral flexion restriction means the left side is the side of the open wedge.

-Right rotational restriction means the spinous has rotated to the right.

-Left rotational restriction means the spinous has rotated to the left.

401. Indication: Left rotation malposition, left lateral flexion malposition. Posterior right, superior spinous

Location: T1

Choose the most appropriate adjustment for this indication from the following list. Also, which is the most correct line of drive?

a. thumb spinous push
b. bilateral hypothenar transverse push
c. modified cervical break
d. bilateral thenar transverse push
e. hypothenar spinous crossed thenar transverse push

f. P-A, L-M
g. M-L, A-P
h. A-P
i. directly posterior
j. M-L

402. Based on the previous question, which is the segmental contact point? Also, identify the contact point.

a. lateral aspect of the right transverse process
b. right lateral aspect of the spinous process
c. medial aspect of the right right transverse process
d. left lateral aspect of the spinous process
e. rib 1
f. distal palmar surface of thumb
g. bilateral thenar
h. pisiform of inferior hand
i. pisiform of superior hand
j. hypothenar

403. Indication: Right rotational malposition, left lateral flexion malposition, right lateral flexion restriction. Posterior left, inferior spinous

Location: C3

Choose the most appropriate adjustment for this indication from the following list. Also, identify the correct patient position and doctor position.

a. thumb spinous push
b. thumb pillar anterior pull
c. index spinous push
d. hypothenar transverse push
e. thenar coastal drop
f. patient is supine
g. patient is prone
h. stand at 45° to the patient at the head of table on the left
i. stand in fencer position on the left
j. stand at 45° to the patient in fencer position on the right

404. According to the previous question, which is the contact point? Also, which is the line of drive?

a. thenar eminence
b. index finger
c. palmer surface of thumb
d. hypothenar
e. pisiform
f. P-A, counterclockwise torque
g. P-A, clockwise torque
h. A-P
g. directly posterior
h. M-L

405. Indication: Right rotational malposition, left lateral flexion malposition.

Posterior left, superior spinous

Location: L3

Choose the most appropriate adjustment for this indication from the following list. Also, identify the most correct line of drive.

a. hypothenar spinous push
b. thenar costal push
c. bilateral mammillary push
d. bilateral thenar mammillary push
e. hypothenar costal push

f. I-S
g. L-M
h. M-L
i. L-M, P-A
j. M-L, P-A

406. Based on question 405, identify the contact point and the segmental contact point.

a. hypothenar of superior hand
b. index finger of superior hand
c. hypothenar of inferior hand
d. index finger of inferior hand
e. thumb of superior hand

f. spinous process
g. mammillary process
h. transverse process
i. articular pillar
j. lamina

407. Indication: Misaligned inferior and anterior

Location: Head of humerus (right glenohumeral joint)

Which is a common segmental contact point to adjust the shoulder related to this indication and location with the patient in a sitting position?

a. clavicle
b. rib 1
c. olecranon process
d. distal ulnar

e. distal radius
f. pisiform
g. sternum
h. scapula

408. Based on the previous question, which is the contact point? Also, identify the correct line of drive.

a. pisiform
b. palmer contact with both hands interlocked
c. index finger
d. thenar
e. hypothenar

f. superior and posterior
g. inferior and anterior
h. anterior only
i. posterior only
j. superior only

409. Indication: Misaligned in anterior direction
Location: Head of humerus (left glenohumeral joint)
Which is a common segmental contact point to adjust the shoulder related to this indication and location with the patient in a supine position?

a. proximal humerus
b. clavicle
c. rib 1
d. rib 2
e. distal ulnar
f. distal radius
g. proximal ulnar
h. sternum

410. Based on the previous question, which is the line of drive? Also, identify the contact point.

a. P-A
b. A-P
c. S-I
d. inferior only
e. superior only
f. clavicle
g. proximal humerus
h. hand and thumbs
i. pisiform
j. thenar

411. Indication: Left lateral flexion restriction, left rotational malposition.
Posterior right, inferior spinous
Location: T8
Choose the most appropriate knee-chest adjustment for this indication from the following list. Also, identify the segmental contact point.

a. bilateral hypothenar transverse push
b. hypothenar spinous push
c. thenar coastal drop
d. index pillar push
e. digit pillar push
f. hypothenar
g. spinous
h. transverse process
i. articular pillar
j. mammillary process

412. Indication: Posterior right, superior spinous.
Left posterior inferior
Location: C7
Choose three appropriate adjustments for this indication from the following list.

a. web costal push
b. thenar coastal drop
c. unilateral hypothenar mammillary push
d. thumb spinous push
e. hypothenar transverse push
f. index spinous push
g. supine thoracic pump handle
h. hypothenar transverse push
i. standing thoracic push
j. hypothenar costal pull

413. Indication: Extension malposition
Location: T4
Choose an appropriate adjustment for this indication from the following list. Also, identify the most correct line of drive.
a. bilateral thenar push
b. hypothenar transverse push
c. thumb spinous push
d. thumb pillar posterior push
e. A-P
f. P-A
g. P-A, S-I
h. P-A, I-S

414. Based on question 413, identify the segmental contact point and patient position.
a. spinous
b. articular pillar
c. transverse processes
d. mammillary processes
e. prone
f. supine
g. side-posture
h. sitting position

415. Based on question 413, identify the direction for the tissue pull. Also, which is the contact point?
a. I-S tissue pull
b. S-I tissue pull
c. bilateral thenars
d. thumb
e. DIP joint of 2nd metacarpal
f. PIP joint of 2nd metacarpal
g. web
h. index finger

416. Indication: Flexion malposition
Location: T2
Choose the most appropriate adjustment from the following list. Also, identify the most correct line of drive.
a. index pillar push
b. thumb pillar posterior push
c. digit pillar pull
d. hypothenar spinous occiput push
e. P-A
f. A-P
g. P-A, S-I
h. P-A, I-S

417. Based on the previous question, which is the segmental contact point? Also, identify the patient position?

a. spinous
b. transverse process
c. mammillary process
d. rib 1
e. prone
f. supine
g. side-posture
h. standing position

418. Based on question 416, which is the contact point?

a. hypothenar of superior hand
b. hypothenar of inferior hand
c. knife-edge of superior hand
d. knife-edge of inferior hand
e. thumb of superior hand
f. thumb of inferior hand
g. web of inferior hand
h. web of superior hand

419. Indication: Flexion malposition
Location: T3

Choose the most appropriate adjustment based on this indication from the following list. Also, what is the most correct line of drive or vector?

a. hypothenar transverse push
b. thumb spinous push
c. bilateral thenar push
d. thumb pillar posterior push
e. P-A, S-I
f. P-A
g. A-P
h. P-A, I-S

420. Based on the previous question, identify the direction of the tissue pull? Also, which is the contact point?

a. S-I tissue pull
b. I-S tissue pull
c. pisiform
d. web of hand
e. tip of index finger
f. bilateral thenars
g. thumb
h. hypothenar

421. Indication: Rib dysfunction
Location: Rib 9

Choose the most appropriate adjustment based on this indication from the following list. Also, identify the contact point.

a. thenar costal drop
b. standing thoracic push
c. hypothenar transverse thoracic push
d. hypothenar coastal pull
e. covered-thumb costal push

f. pisiform
g. thenar eminence
h. bilateral thenars
i. tip of index finger
j. web of hand

422. Based on the previous question, which is the segmental contact point?

a. near rib angle of rib 9
b. T8 transverse process
c. T9 transverse process
d. T7 transverse process

e. T9 spinous process
f. articular pillar
g. lamina of T9
h. rib head of rib 7

423. Indication: Posterior and inferior malposition.
Restricted sacroiliac extension.
Location: Left ilium
Choose the most appropriate adjustment based on this indication from the following list. Also, identify the most correct line of drive.

a. hypothenar pubes
b. hypothenar thigh
c. genu ilium push
d. hypothenar ischium push

e. P-A, M-L, I-S
f. A-P, M-L
g. M-L
h. P-A

424. Based on question 423, which is the patient position? Also, identify the contact point.

a. supine
b. prone
c. side-posture
d. sitting position

e. hypothenar of inferior hand
f. hypothenar of superior hand
g. index finger
h. palmer surface of inferior hand

425. Based on question 423, which is the most correct line of drive? Also, which is the segmental contact point?

a. P-A
b. S-I
c. P-A, I-S, M-L
d. A-P, I-S, M-L

e. ischial tuberosity
f. S1
g. S2
h. PSIS

426. Indication: AI sacrum
Location: Sacrum
Choose the most appropriate adjustment based on the indication from the following list. Also, identify the most correct line of drive.

a. hypothenar sacral apex push — e. A-P
b. hypothenar ischium push — f. P-A
c. external coccyx push — g. M-L
d. hypothenar ischium sacral base push — h. A-P, M-L

427. Based on the previous question, which is the segmental contact point? Also, identify the correct patient position.

a. L5 — e. side-posture
b. L4 — f. supine
c. ilium — g. prone
d. sacral apex — h. sitting position

428. Indication: posterior misalignment
Location: calcaneus
Based on this indication, which is the segmental contact point for adjusting the calcaneus? Also, which is the most correct line of drive?

a. navicular — e. anterior to posterior
b. cuboid — f. posterior to anterior
c. calcaneus — g. superior and posterior
d. fibula — h. superior

429. Indication: anterior misalignment
Location: talus
Choose the most appropriate adjustment based on the indication from the following list. Also, which is the most correct line of drive?

a. P-A glide tibiotalar joint — e. superior
b. A-P glide tibiotalar joint — f. P-A
c. L-M glide tibiotalar joint — g. A-P
d. plantar to dorsal metatarsal — h. L-M

430. Based on the previous question, which is the segmental contact point?

a. calcaneus
b. talus
c. cuboid
d. metatarsal
e. 1st cuneiform
f. 2nd cuneiform
g. fibula
h. tibia

431. Based on question 429, which is the contact point? Also, identify the correct patient position.

a. web of hand
b. pisiform
c. knife-edge
d. thumb
e. prone
f. sitting position
g. side-posture
h. supine

432. Indication: posterior misaligment
Location: talus

Choose the most appropriate adjustment based on the indication from the following list. Also, which is the most correct line of drive?

a. plantar to dorsal metatarsal
b. A-P glide tibiotalar joint
c. P-A glide tibiotalar joint
d. L-M glide tibiotalar joint
e. A-P
f. L-M
g. P-A
h. inferior

433. Based on question 432, which is the segmental contact point?

a. calcaneus
b. 3rd cuneiform
c. 1st cuneiform
d. navicular
e. fibula
f. femur
g. tibia
h. talus

434. Based on question 432, which is the patient position? Also, identify the contact point.

a. prone
b. supine
c. side-posture
d. sitting position
e. standing
f. web of hand
g. thumb
h. index finger
i. pisiform
j. knife-edge

435. Indication: Left rotational malposition, right lateral flexion malposition.
Left posterior superior

Location: T7

Choose the most appropriate adjustment based on the indication from the following list. Also, identify the patient position for this adjustment.

a. thumb spinous push
b. hypothenar spinous occiput push
c. bilateral hypothenar transverse push
d. hypothenar costal pull
e. sitting position
f. supine
g. prone
h. side-posture

436. Based on the previous question, which is the segmental contact point? Also, identify the contact point.

a. spinous
b. transverse processes
c. rib 4 to rib 7
d. occiput
e. bilateral hypothenars
f. web of hand
g. thumb
h. tip of index finger

437. Indication: Right rotational malposition, left lateral flexion malposition.
Right posterior superior
Location: C5

Choose the most appropriate adjustment based on the indication from the following list. Also, identify the patient position for this adjustment.

a. thumb pillar anterior pull
b. supine thoracic pump handle
c. standing thoracic push
d. bilateral mammillary push
e. standing position
f. side-posture
g. prone
h. supine

438. Indication: Posterior right, inferior spinous.
Left rotational malposition, right lateral flexion malposition
Location: C3

Choose the most appropriate adjustment based on the indication from the following list. Also, identify the contact point.

a. bilateral thenar transverse push
b. hypothenar transverse push
c. hyopthenar spinous occiput push
d. index pillar push
e. index finger
f. thenar
g. hypothenar
h. pisiform

439. Indication: Posterior and superior malposition of the sacrum.
PS Sacrum (right)

Location: Right sacrum

Choose the most appropriate adjustment based on the indication from the following list. Also, identify the most correct line of drive.

a. hypothenar ischium push
b. hypothenar sacral base push
c. hypothenar ilium push
d. external coccyx push
e. A-P
f. P-A, I-S
g. M-L
h. inferior

440. Based on the previous question, which is the patient position? Also, identify the doctor position?

a. supine
b. standing position
c. side-posture
d. sitting position
e. standing position
f. fencer stance at 45°
g. square stance
h. sitting posterior to patient

441. Indication: Anterior and superior malposition of the ilium.
AS ilium (left)
Location: left ilium

Choose the most appropriate adjustment based on the indication from the following list. Also, identify the patient position.

a. hypothenar ilium sacral apex push
b. hypothenar pubes
c. hypothenar ilium push
d. hypothenar ischium sacral base push
e. prone
f. supine
g. sitting position
h. side-posture

442. Indication: Lateral misalignment of proximal talus
Location: Talus

Choose the most appropriate adjustment based on the indication from the following list. Also, which is the most correct line of drive?

a. L-M glide-Tibiotalar joint
b. A-P glide-Tibiotalar joint
c. Tarsal Push
d. Tarsal Pull
e. M-L
f. L-M
g. S-I
h. I-S

443. Indication: Lateral misalignment of the tibia.
Location: Tibia

Choose the most appropriate adjustment based on the indication from the following list. Also, which is the most correct line of drive?

a. A-P tibia glide
b. P-A tibia glide
c. L-M tibia glide
d. P-A fibula glide
e. M-L
f. S-I
g. I-S
h. L-M

444. Based on the previous question, which is the segmental contact point?
a. medial portion of tibia
b. lateral portion of tibia
c. medial portion of fibula
d. lateral portion of fibula
e. medial portion femur
f. lateral portion of femur
g. talus
h. calcaneus

445. Based on question 443, which is the patient position? (Choose two answers.)
a. prone
b. side-posture
c. supine
d. sitting position
e. involved hip flexed at 45°
f. involved knee is flexed at 90°
g. involved leg is abducted
h. uninvolved leg is abducted

446. Based on question 443, which of the following is the contact point?
a. thumb
b. web of hand
c. DIP joint of index finger
d. PIP joint of index finger
e. thenar
f. hypothenar
g. middle finger
h. ring finger

447. Indication: Left rotational malposition, left lateral flexion malposition.
Right lateral flexion restriction.
Posterior right, superior spinous
Location: T6
Choose the most appropriate knee-chest adjustment based on the indication from the following list? Also, identify the most correct line of drive.
a. hypothenar spinous push
b. thoracic pump handle
c. hypothenar costal push
d. hypothenar spinous occiput push
e. A-P
f. P-A, L-M
g. I-S
h. M-L

448. Based on the previous question, which is the direction of torque? Also, identify the contact point.

a. counterclockwise
b. clockwise
c. hypothenar
d. thumb
e. web of hand
f. ring finger
g. index finger
h. PIP joint of index finger

449. Based on question 447, which hand is the contact point located? Also, where is the placement of the stabilization or indifferent hand?

a. superior
b. inferior
c. on the opposite transverse process
d. on the same side transverse process
e. anatomical snuffbox of contact hand
f. overlapping the patient's ear
g. on the sacrum
h. over the thumb of contact hand

450. Indication: Left rotational malposition, right lateral flexion malposition.
Posterior right, inferior spinous
Location: T5
Choose the most appropriate adjustment based on the indication from the following list. Also, identify the patient position.

a. hypothenar spinous occiput push
b. unilateral hypothenar transverse push
c. index costal push
d. bilateral index pillar push
e. fencer stance
f. sitting position
g. supine
h. prone

451. Based on question 450, which is the segmental contact point?

a. transverse process
b. spinous process
c. rib 1
d. rib 2
e. rib 3
f. articular pillar
g. mammillary process
h. rib 5

452. Based on question 450, which is the contact point?

a. hypothenar
b. finger tips
c. index finger
d. PIP joint of index finger
e. DIP joint of index finger
f. thumb
g. web of hand
h. thenar

453. Indication: Right lateral flexion restriction, left rotational restriction.
Posterior left, inferior spinous
Location: C4

Choose the most appropriate adjustment based on the indication from the following list. Also, identify the segmental contact point.

a. bilateral index pillar push	e. articular pillar
b. hypothenar spinous push	f. spinous process
c. hypothenar pillar push	g. transverse process
d. thoracic pump handle	h. head of rib one

454. Based on question 453, which is the patient position? Also, identify the position of the stabilization hand or indifferent hand.

a. side-posture	e. across forehead
b. supine	f. across medical clavicle
c. prone	g. across lateral clavicle
d. wrapped around chin	h. across the sternum

455. Indication: Right rotational malposition, right lateral flexion malposition.
Posterior left, superior spinous
Location: L2

Choose the most appropriate adjustment based on the indication from the following list. Identify the patient position.

a. bilateral thenar mammillary push	e. sitting position
b. hypothenar spinous pull	f. standing position
c. hypothenar thigh	g. supine
d. hypothenar ilium-ischium	h. side-posture

456. Based on question 455, which of the following is the contact point?

a. pisiform	e. fingertips of inferior hand
b. thenar	f. fingertips of superior hand
c. hypothenar	g. palmar surface of hand
d. web of hand	h. calcaneal

457. Based on question 455, which is the segmental contact point? Also, posterior left, superior spinous is a Gonstead listing that can also be interpreted as which of the following?

a. spinous process
b. transverse process
c. mammillary process
d. articular pillar

e. PLS-SP
f. PLI-M
g. PRI-M
h. PRS-SP

458. Based on the Gonstead listing in question 457, which is the equivalent National listing?

a. left posterior superior
b. left posterior inferior
c. right posterior superior
d. right posterior inferior

e. left rotational malposition
f. restricted right lateral flexion
g. restricted right rotation
h. right rotational malposition

459. Indication: Left rotational malposition, left lateral flexion malposition.
Posterior right, superior spinous
Location: L1
Choose the most appropriate adjustment based on the indication from the following list. Also, what is the contact point?

a. hypothenar spinous push
b. hypothenar thigh
c. hypothenar pubes
d. bilateral thenar mammillary push

e. hypothenar of superior hand
f. hypothenar of inferior hand
g. palmar contact of cephalad hand
h. palmar contact of caudal hand

460. Based on question 459, which is the segmental contact point? Also, posterior right, superior spinous is a Gonstead listing that can be described as which of the following?

a. femur
b. pubic ramus
c. spinous process
d. mammillary process

e. PRI-M
f. PR
g. PRS-SP
h. PLS-SP

461. Which of the following patient positions can you perform a hypothenar spinous push? (Pick 3 choices.)

a. prone
b. side-posture
c. cervical chair
d. knee-chest

e. supine
f. standing
g. standing with one leg flexed to 90°
h. supine with one leg flexed to 90°

462. Indication: Loss of P-A glide of the humeroulnar joint.
Location: ulna
Choose the most appropriate adjustment based on the indication from the following list. Also, identify the segmental contact point.

a. A-P glide of ulna
b. P-A glide of ulna in extension
c. P-A glide of radial head
d. A-P glide of radial head
e. olecranon process
f. radial head
g. distal ulna
h. humerus

463. Based on question 462, which is the contact point? Pick two choices.

a. pisiform
b. knife-edge
c. web of hand
d. hypothenar
e. thenar
f. thumb
g. index finger
h. palmar contact

464. Indication: Rib1 dysfunction.
Location: Rib 1
Choose the most appropriate adjustment based on the indication from the following list. Also, identify the segmental contact point.

a. thenar coastal drop
b. index costal push
c. covered-thumb push
d. web costal push
e. rib angle
f. medial clavicle
g. lateral clavicle
h. T2 transverse process

465. Based on question 464, which is the most correct line of drive? Also, pick three patient positions in which this adjustment is usually given.

a. A-P
b. P-A, S-I, L-M
c. P-A, S-I, M-L
d. M-L
e. P-A
f. prone
g. supine
h. sitting position
i. knee-chest
j. toggle position

466. Indication: Left posterior inferior.
Posterior right, superior spinous
Location: L4
Choose a knee-chest adjustment that is most appropriate based on the indication. Also, indicate the contact point.

a. hypothenar spinous push
b. reverse lumbar roll
c. covered-thumb push
d. web costal push
e. fingertips of superior hand
f. hypothenar
g. thumb
h. web of hand

467. Indication: Right rotational malposition, left lateral flexion malposition.
Posterior left, inferior spinous
Location: L3
Choose the most appropriate side-posture adjustment based on the indication. Also, posterior left, inferior spinous is a Gonstead listing that can also be described as which of the following?
a. hypothenar thigh
b. thenar costal drop
c. thoracic pump handle
d. hypothenar mammillary push
e. PR
f. PL
g. PRI-M
h. PLI-M

468. Indication: Glenohumeral joint restricted in external rotation
Location: Right Glenohumeral joint
An adjustment is given to correct external rotation. Which would be the segmental contact point for this adjustment?
a. scapula
b. rib 1
c. proximal humerus
d. clavicle
e. proximal ulna
f. distal ulna
g. proxmial radius
h. distal radius

469. Indication: Right rotational malposition, right lateral flexion malposition.
Location: C1
Choose the most appropriate adjustment based on the indication from the following list. Also, identify the contact point.
a. index atlas push
b. calcaneal zygomatic pull
c. occipital lift
d. hypothenar occiput push
e. calcaneal surface
f. thumb
g. index finger
h. hypothenar

470. Indication: Left rotational malposition, right lateral flexion malposition.
Posterior right, superior spinous
Location: T2

Choose the most appropriate adjustment in a patient prone position based on the indication from the following list. Also, identify the doctor position.

a. hypothenar costal pull
b. thumb spinous push
c. thenar costal drop
d. unilateral thenar mammillary push
e. fencer stance
f. square stance
g. sitting position
h. standing position

471. Based on the previous question, which is the contact point?

a. index
b. thumb
c. palmar surface
d. calcaneal surface
e. pisiform
f. middle finger
g. knife-edge
h. PIP joint of index finger

Answers:

1. b, c
2. b, i
3. b, f
4. a, f
5. c, h
6. d
7. d
8. a, g
9. a, h
10. d, e
11. c, f
12. d
13. b
14. d
15. d
16. c, j
17. g
18. a
19. g
20. c, f
21. c
22. g
23. a, i
24. c
25. a
26. e
27. a, b
28. d
29. h
30. a, g
31. b
32. g
33. b, c
34. a, f
35. e
36. c
37. b, e
38. g
39. a, f
40. c
41. a
42. e, h
43. b
44. c
45. d
46. f
47. a
48. d
49. b
50. c
51. h, j
52. a, d
53. b
54. e, f, g
55. a, b, i
56. b
57. c
58. f, g
59. a
60. b, e
61. e
62. a, f
63. c, d
64. d, f
65. g
66. a, c
67. f
68. c
69. e
70. b
71. a, b
72. e
73. f, g, j
74. b
75. c
76. b, f
77. a, c
78. e
79. h
80. a, b
81. f
82. a, e
83. d, f
84. c, i
85. b
86. d, e
87. h
88. d
89. d
90. f
91. d
92. c
93. a, b
94. e
95. f
96. d
97. a, g
98. d
99. c, g
100. b
101. a
102. e
103. d
104. a
105. b
106. g
107. f
108. e
109. d
110. a, b
111. c
112. d
113. h
114. c
115. c, d
116. a, d
117. d
118. a
119. e
120. b
121. a
122. c
123. a, b, f
124. e
125. c, d
126. a, b, c
127. g
128. a, e
129. a
130. c, i
131. d
132. b
133. a, j
134. b, c, g
135. e
136. a
137. b, c, h
138. c, h
139. a, c
140. g
141. b, e, g
142. d, j
143. f, g, h
144. e, g
145. f, g, i
146. d
147. b
148. a, b
149. f
150. e
151. b
152. g
153. a, b
154. c, f
155. f, g
156. h
157. b
158. h
159. a, e
160. a, b
161. e
162. d, e, i
163. f
164. a
165. a, g
166. b, c
167. c
168. d
169. a, h
170. g
171. a
172. b
173. d
174. c, d
175. b, c, h
176. a, b
177. f, h, i
178. f
179. d
180. a, f
181. a
182. d
183. d
184. e
185. a
186. a, c
187. b, c, d
188. a, d
189. h
190. c, g
191. c
192. a, i
193. b, c
194. e
195. c
196. d, f
197. g
198. c
199. e
200. d
201. a, b
202. f
203. a, h
204. e
205. c, g, i
206. c, d
207. e
208. f
209. e, h
210. b
211. g
212. a, b
213. c
214. a, e
215. b
216. g
217. e, h
218. b
219. a, d, g
220. h
221. e
222. b

223. a, b
224. a, f
225. c, d
226. e
227. d
228. a
229. a, e
230. g
231. f
232. d, f
233. b
234. g
235. d
236. c
237. e
238. b
239. e
240. d
241. a, g
242. a, j
243. b
244. g
245. d, h
246. b, f
247. b
248. e
249. d, i, j
250. d
251. a
252. h, j
253. c, d
254. a, b
255. d
256. a
257. g
258. d
259. a, h
260. c
261. e
262. a
263. g, h
264. d
265. c
266. b, f
267. a, b
268. b, d, e
269. c
270. a, h
271. c
272. e
273. h
274. b
275. a, g
276. d
277. e
278. a, b
279. c
280. c
281. d, h
282. c
283. e
284. c, d, g
285. c, f
286. h, i
287. b
288. e
289. a
290. a, j
291. b, f
292. h
293. d
294. d
295. a, g
296. g
297. a, b, c
298. f
299. c, d, i
300. b
301. c, e
302. b, f
303. g
304. f
305. c
306. a, h
307. e, g
308. a, b, j
309. d
310. a, d, e
311. a, f, i
312. c
313. a, i
314. c
315. e, f, h
316. g
317. d
318. d
319. b
320. b, e
321. d
322. h
323. c
324. c
325. b, e, i
326. d, f
327. e
328. a, d
329. g
330. c
331. h
332. f
333. b, f
334. a
335. a, i
336. f
337. d, h
338. c
339. a, f
340. e
341. d
342. c, d
343. a
344. e, f
345. c
346. b
347. e
348. a, j
349. h
350. g, j
351. a
352. e, i
353. d
354. e
355. b, c
356. c, g
357. c, d, i
358. a, h
359. e
360. d
361. c
362. b
363. c
364. f
365. a, b, c
366. e
367. d, g, h
368. f
369. a, g
370. e
371. e, j
372. b, h
373. d
374. h
375. a
376. g, j
377. a, b
378. c
379. a
380. c
381. a, d
382. e, j
383. b
384. d
385. d
386. a
387. c
388. i, j
389. a
390. b, c, e
391. e, h
392. c, f
393. e
394. b, d
395. d
396. f
397. b
398. a, g
399. e
400. c
401. a, f
402. b, f
403. b, f, h
404. c, g
405. a, i
406. c, f
407. c
408. b, f
409. a
410. b, f
411. a, h
412. d, e, f
413. a, g
414. c, e
415. b, c
416. d, h
417. a, e
418. a
419. c, h
420. b, f
421. a, g
422. a
423. c, e
424. b, f
425. c, h
426. a, f
427. d, e
428. c, e
429. b, g
430. b
431. a, h
432. c, g
433. h
434. a, f
435. c g
436. b, e
437. a, h
438. d, e
439. b, f
440. c, f
441. d, e
442. a, f
443. c, h
444. b
445. c, e
446. f
447. a, f
448. b, c
449. a, e
450. b, h

451. a
452. a
453. c, e
454. b, d
455. b, h
456. e
457. a, e
458. d
459. a, f
460. c, g
461. a, b, d
462. b, e
463. f, g
464. b, e
465. b, f, g, h
466. a, f
467. d, h
468. c
469. a, g
470. b, e
471. b